The Ant Trap

OXFORD STUDIES IN PHILOSOPHY OF SCIENCE
General Editor: Paul Humphreys, University of Virginia

Advisory Board
Anouk Barberousse (European Editor)
Robert W. Batterman
Jeremy Butterfield
Peter Galison
Philip Kitcher
Margaret Morrison
James Woodward

The Book of Evidence
Peter Achinstein

Science, Truth, and Democracy
Philip Kitcher

Inconsistency, Asymmetry, and Non-Locality: A Philosophical Investigation of Classical Electrodynamics
Mathias Frisch

The Devil in the Details: Asymptotic Reasoning in Explanation, Reduction, and Emergence
Robert W. Batterman

Science and Partial Truth: A Unitary Approach to Models and Scientific Reasoning
Newton C. A. da Costa and Steven French

Inventing Temperature: Measurement and Scientific Progress
Hasok Chang

The Reign of Relativity: Philosophy in Physics 1915–1925
Thomas Ryckman

Making Things Happen: A Theory of Causal Explanation
James Woodward

Mathematics and Scientific Representation
Christopher Pincock

Simulation and Similarity: Using Models to Understand the World
Michael Weisberg

Systematicity: The Nature of Science
Paul Hoyningen-Huene

Causation and Its Basis in Fundamental Physics
Douglas Kutach

Reconstructing Reality: Models, Mathematics, and Simulations
Margaret Morrison

The Ant Trap: Rebuilding the Foundations of the Social Sciences
Brian Epstein

The Ant Trap

Rebuilding the Foundations of the Social Sciences

BRIAN EPSTEIN

OXFORD
UNIVERSITY PRESS

OXFORD
UNIVERSITY PRESS

Oxford University Press is a department of the University of Oxford. It furthers the University's objective of excellence in research, scholarship, and education by publishing worldwide. Oxford is a registered trade mark of Oxford University Press in the UK and certain other countries.

Published in the United States of America by Oxford University Press
198 Madison Avenue, New York, NY 10016, United States of America.

© Brian Epstein 2015

First issued as an Oxford University Press paperback, 2018

All rights reserved. No part of this publication may be reproduced, stored in a retrieval system, or transmitted, in any form or by any means, without the prior permission in writing of Oxford University Press, or as expressly permitted by law, by license, or under terms agreed with the appropriate reproduction rights organization. Inquiries concerning reproduction outside the scope of the above should be sent to the Rights Department, Oxford University Press, at the address above.

You must not circulate this work in any other form
and you must impose this same condition on any acquirer.

Library of Congress Cataloging-in-Publication Data
Epstein, Brian.
The ant trap: rebuilding the foundations of the social sciences / Brian Epstein.
pages cm.—(Oxford studies in philosophy of science)
Includes bibliographical references and index.
ISBN 978–0–19–938110–4 (hardcover); 978–0–19–087175–8 (paperback)
1. Social sciences—
Philosophy. 2. Social groups. I. Title.
H61.15.E67 2015
300—dc23
2014022916

*For my parents,
with love and gratitude*

CONTENTS

Introduction   1

PART ONE   FOUNDATIONS, OLD AND NEW   11

1. Individualism: A Recipe for Warding off "Spirits"   13
2. Getting to the Consensus View   23
3. Seeds of Doubt   36
4. Another Puzzle: A Competing Consensus   50
5. Tools and Terminology   61
6. Grounding and Anchoring   74
7. Case Study: Laws as Frame Principles   88
8. Two Kinds of Individualism   101
9. Against Conjunctivism   115

PART TWO   GROUPS AND THE FAILURE OF INDIVIDUALISM   129

10. Groups and Constitution   133
11. Simple Facts about Groups   150
12. The Identity of Groups   169
13. Kinds of Groups   182

14. Group Attitudes: Patterns of Grounding   197

15. Group Action: More than Member Action   217

16. Group Intention   236

17. Other Theories I: Social Integrate Models   250

18. Other Theories II: Status Models   264

Looking Ahead   276

Acknowledgments   281
Bibliography   283
Index   293

Introduction

Shortly after I got out of college, back in the early 1990s, I took a job at a management consulting firm. The firm was employed by mammoth companies like Gillette and AT&T, doing projects that now seem almost absurd. A team consisting of several "analysts" like me and a couple of "managers" just out of business school would spend weeks writing questionnaires—*How often do you make international calls? Is a close shave most important to you, or is it more important to avoid razor burn?* Then we would send researchers into malls across the country, stacks of questionnaires in hand. They would survey five to eight hundred people, and our statistics department would type the responses into a computer, analyze them, and send us the results. We would then take those results, sketch out a set of bar charts and scattergraphs, and finally hand them to the production department (in those pre-PowerPoint days) to make a slick presentation.

Companies paid hundreds of thousands of dollars for those presentations. (Sadly, I was a lowly analyst, so I only saw these numbers on the invoices, not in my bank account.) There was a reason they paid so much. They needed information about people—what they buy, what they read, what they do in their spare time, whom they vote for, and how they shave—and there was no other way to get it. These companies had policy decisions to make: what products to develop and what to abandon, which markets to enter and which to flee, whether to hike prices or reduce the length of warranties. In 1993, doing expensive little surveys was the best way to inform such decisions.

The last twenty years have seen a revolution in how we collect data about people. Today, a company does not need to pay the price of a house in Boston to survey 800 mall-walkers. People are throwing information at companies as fast as those companies can collect it. In the next month, 200 million people will run 13 billion searches on Google in the United States alone. In the same period, Facebook will compile personal information from 1.3 billion people across the world. Walmart will process and record 7 billion purchases by 100 million people. Verizon, AT&T, and other wireless carriers will record the locations of 110 billion telephone calls made by 280 million people, and will track the senders and recipients of 170 billion text messages. Nowadays, running a manual survey of 800 people would be both inefficient and unscientific.

Much of this transformation is, of course, explained by technologies: the internet and mobile phones; point-of-sale, tracking, and surveillance systems; data analysis and pattern recognition software; computer processing and storage; and so on. But technology is only part of the explanation. Nearly as important is social change. People have turned out to be surprisingly eager to publicize their personal information. Contemporary labor markets are pushing each of us to advertise ourselves. And the ecosystem of modern corporations has made it an imperative of survival to use personal information in order to increase profits.

Increasingly, economic activity turns on collecting and mobilizing information about people. Industries built for this purpose now dwarf the traditional academic departments and think-tanks that once dominated the social sciences. Google—whose business, after all, is directing people to documents written by people and tailoring advertisements to people—has over 35,000 employees, more than twice the 13,000 academic economists in the United States. And the marketing department of Procter and Gamble is larger than the sociology departments of all US universities combined. It is only a slight exaggeration to say that the world economy is transforming into a massive system for doing social science. For all our talk of the "information economy," the "knowledge economy," and the "technology economy," a more accurate name for the present epoch is the "social sciences economy."

The Paradox of the Social Sciences

Given all this, you would think the social sciences themselves, wielding data that just a few years ago no one had dreamed possible, would be riding high. But despite it all—all the data and all the computers and all the corporate attention—the social sciences are hardly budging. So far, the new advantages have been of little help in deciding among conflicting theories of the workings of the economy, the sources of poverty, the prescriptions for improving education, and financial regulation. If anything, the last few years have deflated whatever optimism we might have had about social theory.

The latest blow came in the form of the recent financial crisis. Just a few years before, economists were gaining confidence in their abilities to understand and guide social systems. In 2004, Ben Bernanke, before becoming chairman of the Federal Reserve, wrote a paper describing the "Great

Moderation" in the global economic system.¹ Like many other economists, Bernanke was impressed by the apparent decline of risk in financial markets, as economies grew less volatile. He saw an end to the successive crises of earlier generations. Bernanke weighed three possible explanations. Perhaps this Great Moderation was a result of structural changes in the economy, such as the shift from manufacturing to services. Perhaps it was a result of improved macroeconomic policies, guided by contemporary economics. Or perhaps it was just good luck.

Of these three possibilities, the second one represents a triumph of applied social science. And this is the explanation Bernanke found evidence to support. "I think it is likely," said Bernanke, "that the policy explanation for the Great Moderation deserves more credit than it has received in the literature."[2] In a classic case of poor timing, Olivier Blanchard, chief economist at the International Monetary Fund, published a paper in early 2008 agreeing with this assessment, saying "The state of macroeconomics is good."[3]

These pronouncements were premature. The unraveling of financial markets in late 2008 took the profession by surprise, its speed and magnitude terrifying economists and policymakers alike. Amid the crisis, the economics profession did not have even roughly consistent recommendations about how to react to it. Prominent economists excoriated the Treasury and the Federal Reserve for every action they took and for every action they failed to take: for allowing Lehman Brothers to fail, for bailing out AIG and protecting its creditors from losses, for pushing an economic stimulus, for effectively nationalizing Fannie Mae and Freddie Mac, for three episodes of quantitative easing, and so on. Inasmuch as the financial authorities deserve some credit—which they surely do—probably the best that can be said is that they did a good job putting out short-term fires, and avoided wholesale catastrophe. But there was no unified theory guiding them.

Since the crisis, economists have been wringing their hands about the discipline. Paul Krugman has been a vocal critic, titling his cover article for the *New York Times Magazine* "How Did Economists Get It So Wrong?"[4] But even the most orthodox voices were shaken. Alan Greenspan, testifying to Congress in 2009, disowned some of his most deeply held beliefs about the rationality of markets: "The whole intellectual edifice," he admitted, "collapsed in the summer of last year."[5] The economist Andrew Lo has reviewed 21 books on

---

[1] Bernanke (2004) 2012.
[2] Bernanke (2004) 2012, 159.
[3] Blanchard 2008. See also Cassidy 2010, Krugman 2009, Kirman 2010.
[4] Krugman 2009.
[5] Alan Greenspan, testifying before the House Committee on Oversight and Government Reform on October 23, 2008.

the financial crisis, and concluded, "there is still significant disagreement as to what the underlying causes of the crisis were, and even less agreement on what to do about it."[6] And Olivier Blanchard has withdrawn his optimism, retracting his earlier views in a paper titled "Rethinking Macroeconomic Policy."[7]

This swing, from unrelenting optimism to self-critical breast-beating, is a familiar story in the social sciences. One doesn't have to be a historian to think of innumerable times that social scientists played the role of Icarus (or Wile E Coyote), thinking they have safely taken flight, only to plunge to earth. Over and over, we have seen plausible theories across the social sciences slapped down.

As compared to past crises, the overconfidence of theorists in 2008 was not extreme. In fact, this is what is depressing about our latest episode. Part of what is noteworthy about the situation today is that, preceding the crisis, the ambitions of social scientists were actually fairly limited. We thought we had learned, through theory and trial and error, not how to create a utopia on earth, not how to solve the world's social ills, but just how to avoid wild economic swings and massive recessions.

Reactions

Many people inside and outside the profession have reacted to the failures of social science as Friedrich Hayek did, back in the 1940s: namely, economies and societies are unimaginably complex systems. As a result, policymakers cannot possibly have enough knowledge to make choices on behalf of a society as a whole. Chances are that they will be worse at it than a distributed market will. Therefore, it is folly even to try to explain, predict, or direct economic activity. In the face of policymaker ignorance, we should minimize policy and regulation, letting the market direct itself rather than trying to give it any direction from above.

A different response to the failures of social science is to be a conservative in the style of Edmund Burke, the eighteenth-century political theorist. Burke too argued that economies and society are too complex to understand. But instead of concluding that we should minimize regulation, he argued that we should be suspicious of abstract reasoning and radical change of any kind. Whatever we do, there is a good chance we will make things worse than they are. On a Burkean approach, it is not the absolute level of regulation that should be minimized, but the pace of change.

---

[6] Lo 2012.
[7] Blanchard et al. 2010, Blanchard 2011.

There is value in both of these reactions. But if the last few years have shown anything, it is that refusing to design and intervene in social systems is often worse than designing them in partial ignorance. Many recent policy failures have been a result of *under*-design, from Donald Rumsfeld's failed "hands off" policy in the Iraq reconstruction to the limits on financial regulation in recent years. Likewise, Burkean conservatism is untenable in many domains. As the economist Paul Romer recently pointed out, if you adopt a set of financial regulations and keep them unchanged, the markets will find a way around them, and ten years later, you'll have a financial crisis.[8] Though they continually disappoint us, theory-led policy interventions—that is, the prescriptions of the social sciences—are indispensable.

What, then, has gone wrong with theories in the social sciences? Why, despite the information revolution, are we not better off? There is no shortage of diagnoses out there. With each failure of the social sciences, theorists have turned their critical sights toward its methods. In many ways, the various methods of the social sciences have been found wanting. The prevailing diagnoses fall, more or less, into five general categories:

(1) *Our models of the individual are inadequate.* Individuals are modeled as rational, when they are not rational. They are modeled as being similar to one another, when they are radically heterogeneous. They are modeled as having perfect information about the world and about the future, and as being perfect calculators of their own interests, when they are far from it. They are modeled as being independently operating atoms, when they are socially constituted. All of these diagnoses criticize the way widely used models treat individual people.[9]

(2) *We have a poor understanding of the "emergence" of group properties out of aggregates of individuals.* Systems of interacting parts often have very different properties than the individuals that compose them. A brain has different properties than individual neurons, an ant colony has different properties than the individual ants, and likewise a society has properties that cannot easily be predicted from the properties of individuals. The diagnosis is that our models of individuals may be ok, but our theories are not good at determining how individuals aggregate into large groups.[10]

(3) *We are building models in the wrong style.* Some theorists hold that our models are too mathematical, or that we have been seduced by the elegance of

---

[8] Romer 2011.
[9] For discussion of a number of these in economics, see Colander 1996.
[10] Approaches to aggregation are omnipresent in the social sciences, drawing on fields such as equilibrium theories, network theory, theories of complex systems, and many others.

certain abstract structures that do not reflect the real world. Others argue that models in the social sciences are not mathematical enough, or use mathematics incorrectly. Still others argue that we will never be able to model society in terms of systems of equations, but that we should perform computer simulations instead.[11]

(4) *We are building models at the wrong level.* From the beginning, the social sciences have been bitterly divided about the right "level" for social explanations. Some theorists argue that macroscopic social phenomena, such as financial bubbles or the growth of economies, can only be explained in terms of other macroscopic social phenomena. Others are committed to explaining social phenomena in terms of individuals. Recently, some theorists have even argued that individuals are too "high-level," and that social theory should be founded in neuroscience.[12]

(5) *"Grand theorizing" is out of our reach altogether.* In recent years, many social scientists have grown suspicious of theories that intend to model societies or economies as a whole. In fact, one of the hottest fields in economics today involves only minimal theory. Instead, it takes its cues from medicine, designing and running randomized trials. Other theorists are devoting their energies to small models that test hypotheses about very narrow parts of the economy.[13]

Different research strategies correspond to each of the prevailing diagnoses. If the rational choice model of the individual is a problem, we should develop more refined theories of individual choice. If the problem is the aggregation of individuals, we should develop mathematical or computational techniques. If the problem is grand theorizing, we should develop experimental methods such as randomized testing.

A Deeper Flaw: The Anthropocentric Picture of the Social World

All of these are plausible diagnoses. To some extent, each of these avenues needs to be explored if we are to make real headway in the social sciences. All

---

[11] For example, in economics see Axtell (2006) 2014; Beed and Kane 1991; Debreu 1991; Epstein 2005; Farmer and Foley 2009; Krugman 2009; Lo and Mueller 2010.

[12] See, for instance, Alexander et al. 1987; Archer 2003; Hoover 2009; Ross 2008.

[13] Mills 1959 and Geertz 1973 are influential critiques of "grand theorizing" in sociology and anthropology. Recent work on randomized trials in economics can be found in Banerjee et al. (2010) 2013; Duflo 2006.

the data and technology in the world only gets us so far if the models that make use of it are flawed. And so it makes sense that legions of theorists, and millions in research dollars, are dedicated to exploring these models and reactions.

In recent years, however, I have begun to worry that much of this effort is misdirected. It is not that the diagnoses are wrong, but that they overlook a deeper problem. The five categories of diagnosis above are not unique to the social sciences. They are diagnoses that one might apply to meteorology, or to cell biology, or to ecology. We might be modeling meteorological phenomena at the wrong level. We might have poor models of the parts of cells. We might misunderstand how ant colonies aggregate out of interacting individual ants.

Implicit in these five diagnoses—and in the practice of the social sciences from its earliest days—is a particular analogy between the social sciences and the natural sciences. Namely, that the objects of the social sciences are built out of individual people much as an ant colony is built out of ants, or a chimpanzee community is built out of chimpanzees, or a cell is built out of organelles.

When we look more closely at the social world, however, this analogy falls apart. We often think of social facts as depending on people, as being created by people, as the actions of people. We think of them as products of the mental processes, intentions, beliefs, habits, and practices of individual people. But none of this is quite right. Research programs in the social sciences are built on a shaky understanding of the most fundamental question of all: *What are the social sciences* **about**? Or, more specifically: *What* **are** *social facts, social objects, and social phenomena*—these things that the social sciences aim to model and explain?

My aim in this book is to take a first step in challenging what has come to be the settled view on these questions. That is, to demonstrate that philosophers and social scientists have an overly *anthropocentric* picture of the social world. How the social world is built is not a mystery, not magical or inscrutable or beyond us. But it turns out to be not nearly as people-centered as is widely assumed.

The term 'anthropocentric' comes, of course, from astronomy. For centuries, astronomers believed that the features of the universe depended in a crucial way on us—on earth and on man. This illusion was natural. Anthropocentric astronomers had perfectly good reasons for believing that the sun, planets, and stars revolved around the earth. Although they ran into problems of prediction and explanation—much like the social scientists of today—they found ingenious ways of patching their theories, for example, the famous Ptolemeic "epicycles." But no refinement of their knowledge of the planets, or the mathematics of orbits, would fix the problems. What was needed was a deeper theoretical revision: they needed to abandon the anthropocentric picture of the universe.

This is a surprising criticism to levy at social science. It is one thing to accuse medieval cosmologists of overestimating the importance of humans in the universe, but quite another to accuse the social sciences of doing so. The phenomena of the social sciences—economic systems, family relationships, education, crime, language—these are things that involve people. How could the *social* sciences be too anthropocentric?

People are not, of course, irrelevant to the social sciences. Social phenomena involve people. The question is how. How exactly are people involved in social facts, objects, and events? How are these things made? What roles do thoughts, actions, and practices play, and how might they fall short?

These are questions about metaphysics. They are questions about the nature of the social world. To make headway on them, we have a number of resources at our fingertips. Metaphysics has, in recent years, become one of the most careful and sophisticated disciplines in philosophy. It has developed and refined a number of tools for thinking about just these kinds of questions. *How does one entity depend on another entity? What are facts, and how are they grounded by other facts?* And so on. Yet few of these tools have been applied in a serious way to the social world.

To be sure, many people in many traditions have theorized about the nature of the social world. From Hobbes to Hume, Comte to Mill, Herder to Durkheim, and Marx to von Mises, theories of the social world abound. The topic is also increasingly prominent in the contemporary philosophical literature. The most influential of these contemporary accounts is John Searle's. In his 1995 book *The Construction of Social Reality*, Searle attempts to give a reasonably comprehensive theory.[14] Others have plunged in as well. Raimo Tuomela has followed up on Searle in several books, detailing more elaborate theories along similar lines.[15] A different approach is taken by Margaret Gilbert in her 1989 book *On Social Facts*. In that book and in a series of subsequent papers, she develops nuanced theories of groups, along with the commitments, norms, and attitudes that accompany group membership. Michael Bratman focuses in particular on the actions and intentions of groups, in his influential account of shared intention.[16] Philip Pettit, in his 1993 book *The Common Mind*, gives a theory of the nature of the social world. And in the 2011 book *Group Agency*, Pettit and his coauthor Christian List give a theory of the nature and actions of groups.[17] Others have also developed theories of institutions, artifacts, and other man-made entities.[18]

---

[14] Searle 1995. He updates the view in Searle 2010.

[15] E.g., Tuomela 2002, 2007.

[16] Bratman 1993.

[17] List and Pettit 2011; Pettit 1993.

[18] E.g., Sally Haslanger, Ruth Millikan, Richard Boyd, Lynne Baker, Amie Thomasson, Crawford Elder, Frank Hindriks, Francesco Guala, Ron Mallon, and others.

It is not, in other words, that social metaphysics—that is, the nature of the social world—has escaped attention. Yet only recently have people really started to examine the metaphysics in detail. Historically, questions about the nature of the social world were treated in a fairly cursory way, dispatched quickly to make way for points about morality or politics or game theory. And so the sophisticated toolkit of metaphysics mostly sat idly by. Because of this neglect, the settled view of the social world has gone more or less unchallenged.

Social Metaphysics and Social Groups

If it is true that we misunderstand the building blocks of the social world, it is no surprise that we are having trouble in the social sciences, since that misunderstanding distorts our models. Some of the most obvious cases are financial markets. Despite their prominence in the daily newspaper, just what financial markets and financial instruments are, or what their function is, has never been clear to economists. And so they have largely been left out of models, particularly models in macroeconomics. Economists have rationalizations for this: at least until 2008, it was common to argue that the "financial economy" does not bear too much on the "real economy" of houses, cars, and dish soap.[19] In recent years, that argument has fallen flat, and economists have been scrambling to figure out how financial markets and instruments should figure into macroeconomic models. But that scramble does not change the basic problem. Knowing that we need to incorporate financial markets and instruments into our models does not help much if we are clueless about their building blocks. Until we improve our understanding of their nature, we do not have a prayer of modeling them well.

While the exclusion of financial markets from macroeconomics is a glaring example, it is far from the only case of a distorted understanding of the social world. In fact, the field of social metaphysics is only in its infancy. Our flawed understanding starts with much simpler things than the financial economy. Even the very simplest cases are thornier than one might imagine.

A prime example of a simple case is a *group of people*. The social world is rich with groups: classes, populaces, mobs, legislatures, courts, faculties, student bodies, and so on. In any social science, we are interested in investigating facts about groups, facts like the educational attainments of kindergarteners, the voting patterns of legislators, the levels of corruption in bureaucracy, the responsibilities of soldiers for the conduct of war, or the salaries of university

---

[19] See, for instance, Kydland and Prescott 1982; Lucas 1972, 1977.

faculties. And, it seems, the building blocks of groups couldn't be any simpler. A group of people is constituted by people, no more, no less.

But this apparent simplicity is deceptive. A close look at the metaphysics of social groups shows it to be subtler than this. One trick is in the word "constituted." As I will discuss later on, it is technically true that groups of people are *constituted by* people. Constitution, however, has received an enormous amount of attention in the recent metaphysics literature. In the last few years, it has become clear that to say "x is constituted by such-and-such" only gives a tiny bit of information about what x is. It is not hard to see this. One of the examples I will be discussing in some detail is the United States Supreme Court. It is small—nine members—and very familiar, so there are lots of facts about it we can easily consider. Even a moment's reflection is enough to see that a great many facts about the Supreme Court depend on much more than those nine people. The powers of the Supreme Court are not determined by the nine justices, nor do the nine justices even determine who the members of the Supreme Court are. Even more basic, the very existence of the Supreme Court is not determined by those nine people. In all, knowing all kinds of things about the people that constitute the Supreme Court gives us very little information about what that group is, or about even the most basic facts about that group.

These quick observations about the Supreme Court raise more questions than they answer. But that, for now, is the point. Even to understand the nature of simple social groups, we need to take the metaphysics seriously. This book is written with the conviction that we are wasting our time with the most complex cases, if we get even the simple ones wrong.

Part One of this book sets out a general framework for social metaphysics. How do we approach the problems of social metaphysics, what are the projects involved, what are the tools we need, and why have people gotten it so wrong? Part Two applies the tools of social metaphysics to groups. Groups are not even close to being the only social entity. But they are important in their own right, and figuring out how to work with them gives us a template for approaching more complicated things. Groups are also a powerful example for advancing the central point of the book. My aim is to allow us to start freeing ourselves from "the ant trap"—the anthropocentric picture of the social world as being composed by individual people. For this aim, inquiry into groups strikes the target directly. If anything in the social world should be anthropocentric, it is groups of people. Even the most lukewarm defender of anthropocentrism may find it hard to see what could possibly be wrong with an anthropocentric picture of groups. Thus when, in Part Two, we see that anthropocentrism is wrong even for groups, we plant the stake deep into its heart.

PART ONE

FOUNDATIONS, OLD AND NEW

1

Individualism: A Recipe for Warding off "Spirits"

In the middle of the twentieth century, a battle raged between "individualists" and "holists" about the nature of the social world. At the center of these battles was the philosopher Karl Popper. Popper is best known, nowadays, for his work on scientific hypotheses and how they may be falsified. But in his heyday, he was much more famous for his book *The Open Society and its Enemies*, one of the bestsellers of 1945. In that book, he railed against what he took to be a cancer in social theory: the idea that societies have collective minds that direct their activities. *The Open Society* describes a menagerie of philosophical villains, all supposed opponents of human freedom. Karl Marx was probably the most wicked, but following close behind was G. W. F. Hegel.

Popper ascribes to Hegel a disturbing view of political systems. He argues that Hegel privileged states, like Germany, Britain, and the Roman Empire, over individuals. Hegel believed, according to Popper, that the interests of states are more fundamental than those of individuals, and that the lives of individuals are in service to the state, rather than the other way around. Behind this, says Popper, was a twisted metaphysics of the state: the idea that states are autonomous, thinking organisms, something above and beyond the people that comprise them. Instead of taking states to be just aggregates of people, this metaphysics ascribes conscious "spirits" to states:

> The collectivist Hegel . . . visualizes the state as an organism . . . Hegel furnishes it with a conscious and thinking essence, its "reason" or "Spirit." This Spirit, whose "very essence is activity" . . . is at the same time the collective Spirit of the Nation that forms the state.[1]

---

[1] Popper 1945, 41.

By Popper's account, the trouble begins with the distorted idea that the state is a real thing or substance, a whole that exists separately from the people that comprise it. Hegel's first error, he argues, was to "reify" the state, treating it as a real and separate thing. Once that mystical metaphysics was underway, it became natural to see the state as having interests and values of its own.

It was clever for Popper to choose Hegel as an enemy. The link between Hegel and Marx made Popper's book a bible for anticommunists, whose ranks were swelling in the late 1940s. It did not matter much that Popper's depiction of Hegel was a caricature. Philosophers nowadays do not look kindly on Popper's scholarship. Even so, we have to give Popper credit for capturing a genuine anxiety of the time, and also a genuine philosophical problem. The issue Popper gives voice to—that many social scientists make use of a worrisome metaphysics—makes sense. Popper was right that social science has found it hard to avoid speaking of society's "spirit," and that it was never particularly clear what that spirit was supposed to be. Over the years, many theorists struggled unsuccessfully to avoid social "spirits." I will mention two, one from the nineteenth century and one from the twentieth.

1. Leopold von Ranke was arguably the premier historian of the nineteenth century. Ranke was the first truly modern historian, pioneering rigorous historical research. Moreover, he was a fierce opponent of Hegel. He rejected the idea that you could tell a single unified history of the world. The histories of different societies and times, he argued, do not belong in one narrative. And, to be sure, history does not unfold in any particular direction. Ranke did admit the possibility of moral and cultural progress, but he did not see a logic to the history of the world. Instead, he professionalized history, transforming it into a specialized field based on empirical investigation. The job of a historian, according to Ranke, is to craft explanations from the detailed investigation of historical particulars in their contexts.

Despite this, Ranke could not figure out how to elude "spirits" in historical explanation. Social epochs, for Ranke, are marked by particular tendencies, and tendencies differ from epoch to epoch. The goal of social research is to uncover those tendencies, as they manifest in particular events. Together, these tendencies reflect a state's spirit, which is the driving force of human history, much as a person's soul animates his or her life:

> If we now ask, "What is it which enables a state to live?" then it is the same as with the individual, whose life incorporates both body and spirit. So too with the state. Everything depends on spirit, which is the pre-eminent of the two.[2]

---

[2] Ranke (1836) 1981, 112.

Even the scientifically minded Ranke was drawn into this position. Nations, states, and societies perform actions. They go to war, they are party to trade deals, they rise and fall. These actions are not coincidences, but occur at least in part as a result of the general attributes of the nations, states, and societies in question. Ranke was interested in explaining such historical events. To do so, he appealed to tendencies, but then needed something to account for those tendencies. And so, despite disagreeing with Hegel on so much else, he had little choice but to appeal to "spirit."

Ranke was not particularly happy about this. The work of history, he argued, is not the "speculation of philosophers." Historical terms like 'tendencies' were empirically grounded, derived from the investigation of facts, as distinct from abstract concepts. Still, he was aware that even though he denied the universal progress of historical epochs, something was required to unify any given epoch and its tendencies. And so the awkward metaphysics of spirits remained.

2. A second illustration comes from Harvard sociologist Talcott Parsons, a figure from the dominant school of sociology in the middle of the twentieth century. In his work, Parsons observed that the actions we take as individuals—whether we are mothers, soldiers, writers, or fraternity brothers—are performed within a cultural structure. Consider a fraternity, for instance. When a fraternity brother hazes new pledges, he is participating in a system largely dictated to him. He does not choose, but rather is pressed into, the hazing traditions. When the brother has some "creative" idea, like making new pledges stand naked in a snowstorm, this creativity is really a minor variant on a prescribed theme.

The key to explaining the member's action—that is, his sending the freshmen outdoors—is to explain why the hazing system is in place. Here Parsons has a straightforward answer. Hazing serves a "pattern maintenance" function for the fraternity.[3] Its function might be to bind members to one another through common hardship. Or it might be to make the fraternity more attractive and exclusive by creating artificial barriers to entry. In either case, his action helps maintain the fraternity's patterns of behavior. In performing the ritual, the fraternity brother is playing a role in this structure. The "structure," then, is a powerful tool for explaining why he does what he does.

Of course, "structure" is not the same thing as "spirit." By the time Parsons was writing, the social theories of the nineteenth century had been enormously improved upon. Even by Ranke's time, social theory had disposed of a number of worrisome commitments, and by the end of the nineteenth century, many of the assumptions Ranke had made were

---

[3] Parsons 1951, 1954.

also abandoned. For example, Ranke did not manage to shake off the idea that the agents moving the course of history were nations or states, acting something like organisms, with their own unified interests and directions. In the late nineteenth century, theorists gave up on these commitments. Moreover, the religious overtones of social progress had withered as science had matured over the nineteenth century. Even the focus of social science on states themselves had faded with the rise of theories of class interests. By 1900, the term 'spirit' had fallen from fashion in the social sciences.

Yet even in the middle of the twentieth century, when Popper and Parsons were both writing, the lurking presence of some incomprehensible ectoplasm lived on. Parsons did not talk of "spirit," but nonetheless he did not manage to find metaphysically secure ground, free of mystical social unities. To many theorists, Parsons's theory is as imbued with mysticism as Ranke's, especially in its apparent suggestion that societies somehow direct, or are autonomous from, individual action. For instance, the sociologist Harold Garfinkel, in a biting critique, accused the theory of insulting individual autonomy and individual intelligence. Every day, as individuals we talk with our friends, families, and therapists about how we should live our lives, Garfinkel pointed out. We seem to be guided by our own goals, thoughts, and imaginings. But Parsons's theory seems to assign genuine agency to social structures, not to individual people. Garfinkel argues that Parsons's theory portrays us as little more than puppets, who play our role as dictated by society. As he famously put it, the theory turns people into "judgmental dopes."[4]

Much like Ranke, Parsons found it indispensable, when giving social explanations, to appeal to some kind of social unity that was different or separate from individual people. The reasons for this are clear: how we act and what we do differs radically from one society to another, and from one time to another. These differences are a product of the social contexts in which we are embedded. Those social contexts have properties, they change, they affect us. When we talk about them, we are talking about *something*. Both Ranke and Parsons seem to have "reified" the social world in order to construct explanations about social phenomena. Ranke found that historical explanations were best given in terms of social tendencies, and Parsons held that explanations for individual action were best given in terms of the social functions that those actions played. Was this sort of move sound, or was it mystical, as Popper argued?

---

[4] Garfinkel 1967.

The Trouble with Social Entities

This debate is an instance of a familiar pattern in contemporary philosophy. It does not arise just in connection with individuals and society. Rather, it is one example of a common problem in *interlevel metaphysics:* the discipline which studies the nature of "high-level," or "macroscopic," things and how they are related to "low-level," or "microscopic," things. (*Macroscopic* and *microscopic* contrast things at intuitively different levels of organization: for example, to contrast macroscopic things like economies and governments [the things treated in *macro*economics] with microscopic things like individuals, households, and firms [the things treated in *micro*economics]. Or to contrast macroscopic things like climates and oceans, with microscopic things like clouds and waves. Or things like proteins and strands of DNA, with things like protons and electrons.[5])

Is there anything to society above and beyond individuals? The structure of this problem is similar to that of other problems in interlevel metaphysics. We see it in the relation between biology and chemistry: is there anything to life, or to living organisms, above and beyond chemicals? Or in the relation between minds and brains: is there anything to the mind, or to thinking, above and beyond the firing of neurons?

In each of these domains, there seem to be two different stances one could take. First, there is the *dualist* stance: Yes, there is something to living things over and above interacting chemicals, some "vital essence." Yes, there is something to the mind, over and above interacting neurons, some "soul" or "thinking substance." Yes, there is something to society over and above interacting individuals, some "spirit" or "social substance."

Or there is the *monist* stance: No, there is nothing to living things, over and above interacting chemicals; nothing to the mind, over and above interacting neurons; nothing to society, over and above interacting individuals.

In the philosophy of mind, dualism is most closely associated with Descartes, who distinguished two different kinds of substance: mental substance and "corporeal" or physical substance. Corporeal substance operates according to mechanical rules. And according to Descartes, if there were only corporeal substance, there could be no subjectivity and no thinking: the physical character of the brain alone is not enough to determine the properties of the mind. As soon as Descartes introduced his dualism, nearly 500 years ago, it came under attack, and the attacks have continued more or less nonstop.

---

[5] It is problematic even to divide the sciences into "levels." See, for instance, Thalos 2013; Wimsatt 1976, 1994.

Even in Descartes's own day, it was clear that mental substances raise as many problems as they solve. For instance:

1. How can physical matter be guided by physical laws, and yet leave room for the mental to intercede in the course of events?
2. If the mental and physical are separate substances, what is the mechanism for the mental to interact with the physical?
3. How are minds individuated? What makes a mind what it is, and what distinguishes one mind from another?

These sorts of difficulties make it much less tempting to solve the problem by postulating the existence of some kind of separate mental substance.

As alternatives to dualism, philosophers have developed a range of "monist" theories of the mind. In these theories, there is only one kind of stuff—physical stuff—and mental facts are exhaustively determined by physical facts. The relation between physical facts and mental ones can be quite complicated. But in a sense, there is nothing more to the mental "over and above" the physical.

The idea that societies and other social entities are separate and autonomous substances is thus just one species of dualism. And dualism in social theory has the very same problems as it does in theories of mind. If the dynamics of economics, politics, and history proceed through the actions of individuals, how can social laws have any influence on human affairs? Even if they could, by what mechanisms would such things as "structures" influence individuals? And how can we tell what "structures" are in the first place? Despite the considerations motivating Ranke, Parsons, and their ilk, it seems like an enormous step backward for us to reify the social world.

Methodological Individualism as the Alternative

If "reifying" social entities amounts to mysticism, what other choice do we have in understanding them? The alternative that developed in the long-running battle is "individualism." Put very roughly (with the promise of refining it as we go), this is the view that the social world is made up of nothing more than individual people. In *The Open Society*, Popper approvingly cites John Stuart Mill, who insists that social phenomena are nothing more than the thoughts and actions of individual people:

> Thus "all phenomena of society are phenomena of human nature," as Mill said; and "the Laws of the phenomena of society are, and can be,

nothing but the laws of the actions and passions of human beings," that is to say, "the laws of individual human nature. Men are not, when brought together, converted into another kind of substance."[6]

Popper does not follow Mill slavishly. In fact, much of his book is dedicated to criticizing Mill's view that we should fold social theory into the discipline of psychology. But Popper does endorse Mill's thinking about what society *is*. There is nothing more to facts about society, above and beyond facts about individual psychology.

One of Popper's students, J. W. N. Watkins, became the dominant voice in favor of individualism in the mid-1950s. Watkins elaborates the claims of individualism in much more detail than Popper did. One of his sharpest polemics is against the sociology of Talcott Parsons. Watkins accuses Parsons of taking a view that is basically theological, and just layering a secular veneer on it. According to Watkins, Parsons does not distinguish his view much from the view that history is guided by divine providence. After all, Parsons does not object to the idea that the actions of individuals are guided by some social entity. All Parsons does, says Watkins, is replace the divine plan with something that seems more scientific. Parsons, like Hegel, is a "methodological holist":

> On this view, the social behavior of individuals should be explained in terms of their positions within its cultural-institutional structure, together with the laws which govern the system . . . This is what is called methodological holism.[7]

In contrast, Watkins proposes "methodological individualism":[8]

> It is people who determine history, however people themselves are determined. This factual or metaphysical claim has the methodological implication that large-scale social phenomena like inflation, political revolutions, etc., should be explained in terms of the situations, dispositions and beliefs of *individuals*. This is what is called methodological individualism.[9]

---

[6] Popper 1945, 101. Though Popper endorses the metaphysics of "psychologism," he does not endorse it as a theory of social explanation. See Gellner 1973; Udehn 2001, 2002; Wisdom 1970.
[7] Watkins 1955, 179–80.
[8] This term was coined by Joseph Schumpeter in 1908, but only gained currency later on.
[9] Watkins 1955, 179–80.

Worries about dualism remain, Watkins points out, even if we replace talk of social spirits with social structures, and even if we moderate theories like Parsons's, so that people are not quite so dopey. So long as we talk about social structures and social functions and cultural systems, about bases and superstructures and frameworks of oppression, even about nations and institutions and corporations, we risk treating them as if they are real objects or agents, with intentions, plans and goals, and governed by their own laws or logic. From a metaphysical perspective, all these seem questionable, unlike the more sensible view of those who take the social world to be nothing more than individual people.

As the name suggests, methodological individualism is a view about the proper *methodology* of the social sciences. Methodological individualists take a certain attitude toward theories, explanations, or models, in the social sciences. They argue that these are best given in terms of individual people: a "holist" theory, like Parsons's, is a bad theory, a poor explanation.

In the 1950s, debate grew feverish between methodological individualists and methodological holists. In that period, a number of promising individualistically inclined theories flowered, with some especially exciting developments in economics.[10] And individualists like Popper and Watkins were persuasive, not only in their insistence that holist theories were predicated on a mystical metaphysics, but also in their claims that holism threatened individual freedom.

On the other hand, there were plenty of social theorists who seemed to be giving useful and adequate explanations without putting them in terms of individuals. Huge projects in macroeconomics, for instance, were underway at the time, gathering and modeling aggregate measures of economies as a whole.[11] And holists started coming up with examples of phenomena that seemed particularly resistant to individualistic explanations. In a 1955 article, the philosopher Maurice Mandelbaum described the case of an individual withdrawing money from a bank teller window.[12] The behavior of the bank teller and the customer, Mandelbaum points out, depends on certain socially defined roles they play, in the circumstances they are in. It is impossible, he argues, to dispense with "societal facts" in explaining their actions. Such were the holist arguments against individualists, to which the individualists of course had further replies.

---

[10] Particularly important were game theory (Nash 1950; von Neumann and Morgenstern 1944) and general equilibrium theory (Arrow and Debreu 1954).

[11] E.g., Klein and Goldberger 1955; Tinbergen 1956.

[12] Mandelbaum 1955, 308–9.

It was only very late in the course of these debates—once they had already started to run out of steam—that philosophers started to realize the deep confusion at their heart. Neither side had taken one distinction seriously enough: the distinction between individualism about the *metaphysics* of the social world, and individualism about *explanations* of social theories.

Explanatory Individualism vs. Ontological Individualism

It took until 1968 for people to notice and start untangling this confusion.[13] In a paper titled "Methodological Individualism Reconsidered," Steven Lukes inserted a wedge between two different theses that had routinely been conflated in earlier debates.[14] Methodological individualism, Lukes pointed out, was not just a single thesis: it was two. In fact, it consisted of one controversial thesis and one trivial thesis. The controversial part has become known as *explanatory individualism*, and the trivial part as *ontological individualism*.

Explanatory individualism is a thesis about social explanation. It is the claim that social facts are best *explained* in terms of individuals and their interactions. That is, theories and models in the social sciences should be individualistic. They should model the properties of individual people and the interactions among individual people.

This, however, is a separate thesis from ontological individualism. Ontological individualism is a thesis about the makeup of the social world. It holds that social facts are *exhaustively determined* by facts about individuals and their interactions. Ontological individualism says nothing about theories or models or how best to construct explanations in the social sciences. All it says is that there is nothing more to societies, their composition and their properties, above and beyond individual people. Explanatory individualism is a stronger thesis than ontological individualism. Even if societies consisted of nothing more than people, it may be impractical or impossible to construct social explanations individualistically. Even if explanatory individualism is false, in other words, ontological individualism need not be.

Lukes, for his part, rejects the need for (and often, the possibility of) explaining social phenomena in terms of individuals. But he accepts the ontological thesis that

---

[13] Goldstein 1958 noticed that these needed to be pried apart. Unfortunately, he got tangled in a long discussion of ideal types and "anonymous individuals," and failed to insert a wedge between the two theses. This was only exacerbated in a fruitless multi-part exchange between Goldstein and Watkins in the *British Journal for the Philosophy of Science*.

[14] Lukes 1968.

social phenomena are fully made up of individualistic ones. In other words, he sides with Parsons and Mandelbaum on explanation, while agreeing with Popper and Watkins on ontology. Lukes begins with a statement of individualistic ontology:

> Let us begin with a set of truisms. Society consists of people. Groups consist of people. Institutions consist of people plus rules and roles. Rules are followed (or alternatively not followed) by people and roles are filled by people. Also there are traditions, customs, ideologies, kinship systems, languages: these are ways people act, think and talk. At the risk of pomposity, these truisms may be said to constitute a theory (let us call it "Truistic Social Atomism") made up of banal propositions about the world that are analytically true, i.e. in virtue of the meaning of words.[15]

Lukes's argument thus begins with a concession about the nature of social facts, about what they "consist of." He points out, however, that this does not imply that explanations can be given individualistically. He notes that there are many forms of explanation, among which there are perfectly good ones that do not involve individuals at all. And Lukes uses Mandelbaum-style examples to show that in many cases, we should not expect to be able to give strictly individualistic explanations of many social phenomena. Even if facts about bank withdrawals are exhaustively determined by facts about large numbers of individual people, that does not mean that we can construct individualistic explanations of banking.

In the minds of most philosophers, the distinction between explanatory individualism and ontological individualism divides the uncontroversial issues from those that are to be debated. The ontological thesis is settled: there is no more to society, over and above individual people. What remains open is how best to construct social theories and explanations. Are the best social explanations individualistic? Once we sever the link between ontological individualism and explanatory individualism, we can endorse the former and debate the latter.

Having separated these two theses—a controversial thesis about explanation and an uncontroversial one about ontology—philosophers were on their way to a consensus about dualism in social theory: There is no need to be a dualist. Instead, we can be ontological individualists, and still debate explanatory individualism.

They were on their way, but still not quite there. The Lukes paper is not particularly precise. He paints the distinction between ontological and explanatory individualism only in broad brushstrokes. For a time, this was good enough for social theorists, but it soon became clear that we could do better.

---

[15] Lukes 1968, 120.

2

Getting to the Consensus View

For thirty or forty years now, ontological individualism has quieted our fears of social "spirits." All sides seem to agree: *We do not need to worry about the ontology of the social world. The social world is nothing but people and their interactions. Of course, we can still fight about methodology: ontological individualism does not imply explanatory individualism. Maybe explanations should be individualistic, maybe not. But the ontology is safe.*

What, exactly, does it mean to endorse ontological individualism? How do theorists currently understand this claim? In the last generation, a broad consensus has developed, built on the shoulders of two other fields in philosophy—the philosophy of science and the philosophy of mind. From the 1960s to the 1980s, philosophers in those fields made great strides in understanding the relations between macro and micro levels.

Three advances in particular merit attention. First is the idea of *theory reduction*: what does it mean to "reduce" a theory at a macro level to one at a micro level? Second is the idea of *reduction failure*: why might it be impossible to reduce a macro-level theory to a micro-level one? And third is the idea of *supervenience*: even if we cannot reduce a macro-level theory to a micro-level theory, what does it mean to say that there is nothing to the macro level "over and above" the micro level?

These three advances allowed philosophers of social science to move far beyond Lukes's vague distinction. The contemporary consensus is this: ontological individualism should be understood as a claim about supervenience, and this claim is obviously true. In chapter 3, I will raise doubts about these conclusions, but first we must understand this claim and the key advances that led to it.

In discussing this material, I have an ulterior motive. As soon as explanatory individualism is contrasted with ontological individualism, many people fall back on a tempting view. Namely, that social facts are "emergent," that they emerge from individuals and the interactions among individuals. With this chapter, I want to underline that emergentism is little different from the prevailing, consensus view. When, in subsequent chapters, I challenge the consensus view, I also challenge emergentism. This book is just as much a challenge to the theory that society "emerges" from individuals and their interactions as it is to other versions of the consensus view.

Reduction

Let's begin with the idea of a "reductive" explanation. Consider the relation between "macro" and "micro" phenomena in a different sort of system. Consider the behaviors of a school of fish—a huge school of herring, for instance.

Atlantic herring live out most of the year in the North Atlantic, feasting on microscopic creatures. In early winter, these creatures sink into the depths of the ocean, so billions of herring migrate to the Norwegian fjords. (They stay there until January, when they turn back across the Atlantic to spawn.) The fjords are about as protected a place as the herring can find, but their stay is not altogether peaceful. Where herring gather, pods of killer whales follow, attacking in coordinated multiflank maneuvers. To stay alive, the herring need strategies to protect themselves.

In November 1993, the biologists Leif Nøttestad and Bjørn Axelsen sent a small boat out into the fjords to study these strategies. Using a sonar imaging system, they tracked schools of herring as they responded to whale attacks. A single herring school is enormous, consisting of as many as 50 million fish, and stretching for a quarter mile. Imaging entire schools, Nøttestad and Axelsen found that a school behaves as if with one unified mind when whales approach.[1]

When a whale swims at it, a school may split in two, half the fish going left and half going right. Or the school may take a sharp turn in one direction. Or, if it is a particularly large school, it may create a moving vacuole around the whale as it passes through, the herring gathering in propagating waves of density. Or the fish may cluster together in a tight defensive ball, herring packed shoulder to shoulder as if they were already in a jar of cream sauce. A school of herring exhibits a wide range of "defensive" patterns, parrying in clever ways with each thrust of the predators. With these countermeasures, the herring can largely escape mass slaughter.

How do the herring manage this melee, knowing as a group whether to cluster, or split in two directions, or rapidly reverse? What makes the fish at

---

[1] Nøttestad and Axelsen 1999.

the edges of the ball join the cluster, where they are most likely to be eaten, rather than swim off on their own? What makes them so apparently altruistic, sacrificing themselves for the safety of the group?

Back in 1971, the biologist W. D. Hamilton had offered a rather depressing hypothesis to answer questions like these. In "Geometry for the Selfish Herd," he argued that the members of a herd need not be altruistic for the herd to exhibit apparently coordinated defenses.[2] A number of theorists in the 1950s and 1960s had used schooling behavior to defend theories of "group selection"—theories in which the evolution of individual traits is explained by the advantages they provide to the group of those individuals.[3] They argued that the gregariousness of individuals, clustering as they do under threat, is mutually protective for the group. Hamilton argued for a simpler mechanism. When a member of a herd or school is threatened, Hamilton suggested, it does one simple thing. It tries to hide. Unfortunately, the only thing for a herring to hide behind is another herring. So every herring tries to dart behind the others. And as they do so, the packing of the school as a whole generates patterns, such as splitting in two, creating a vacuole, or forming a ball. One simple behavior generates a variety of macroscopic patterns in the aggregate.

To evaluate this hypothesis, Nøttestad and Axelsen used their sonar to record detailed measurements of the whales and the schools of herring. They recorded the reactions of the schools and the countermeasures they took under what circumstances, comparing the results with computer simulations. To the dismay of many an altruist, they found that Hamilton's theory was all they needed. No self-sacrifice, no coordination mechanisms, no instinct to cluster. Just a swarm of selfish herring, acting according to a simple rule. All the complex behaviors of the school "emerged" from a large number of individuals interacting with one another.

This sort of theory, should it turn out to be correct, allows us to talk about schools of fish and their properties, without taking on any sort of dualism about schools. A Hamilton-style explanation allows us to speak of the school's "strategies" without worrying that we are speaking of some independent metaphysical realm. The strategies become no more than a shorthand or abbreviation for the complex of individuals that comprise them. This kind of explanation is a "reduction" of the macro-level theory to the micro-level theory.

---

[2] Hamilton 1971.
[3] E.g., Wynne-Edwards 1962; Pitcher and Parrish 1993.

Humans, Insects, and Antidualism

When it comes to human societies, we tend to see ourselves less often as schools of fish than as swarms of bugs. Homer speaks of the Achaeans buzzing wildly like bees around their warships, and Plato takes bees as a moral model of organization and industriousness. Bernard Mandeville uses insects to illustrate the aggregation of sinful individuals into a virtuous society, while Thoreau compares industrial society to the breeding of workers in the abdomen of an ant queen.[4] The nineteenth-century philosopher Herbert Spencer argued that both ant colonies and human societies were instances of "superorganisms." (This view is coming back into vogue in some circles.) And C. K. Ogden, the translator of Wittgenstein's *Tractatus*, built his nominalist theory of language on the lessons of another book he translated, Forel's *Social World of the Ants*.[5]

Many aggregates have different properties from those of the individuals that compose them: a pool of water has waviness, viscosity, and so on, while these properties do not apply to individual water molecules. This is not always so obvious in the social sciences. Social scientists are fond of pointing out what we might call "aggregation reversals," that is, when a group of people displays the seemingly opposite property that was displayed by the individuals composing it. John Maynard Keynes, for instance, argued for the "paradox of thrift," where an increase in savings by a large number of individuals in an economy can causally produce a reduction in the aggregate savings of the economy as a whole.[6] Economists are also fond of quoting Adam Smith's famous statement, "It is not from the benevolence of the butcher, the brewer, or the baker, that we can expect our dinner, but from their regard to their own interest."[7] Public welfare, it is suggested, can arise from private selfishness. Mandeville, in his essay *The Fable of the Bees*, makes the point even more vividly. Where there is a class of vain and lazy nobles, there is demand for herdsmen, weavers, tailors, furriers, and more. If these vices of vanity and laziness were replaced by honesty, temperance, and toil, we might live in a society with greater individual virtue, but collapsing prosperity. In Mandeville's view, the "diseases" of lust, sloth, avarice, and pride in individuals are essential for a healthy society.[8]

---

[4] Lattimore 1951, 84–96; Plato 1969, 267–9; Mandeville (1714) 1934; Thoreau (1854) 1966.
[5] Spencer 1895; Forel 1928; Heims 1993; Sleigh 2007, 141–50.
[6] Keynes 1936.
[7] Smith (1776) 2006.
[8] Mandeville (1714) 1934.

This echoes Hamilton's "Geometry of the Selfish Herd," where selfishness at an individual level produces gregariousness in the herd.[9] It is a striking fact that properties of individuals can produce their reverse in the aggregate. But the real insight of explanations like these is not that group benefits arise from individual selfishness. Rather, it is that complex and varied properties of aggregates can arise from simple properties of their members in combination with one another.

Looking down at swarms of pedestrians from atop a skyscraper or at troops of ants from a picnic blanket, we perceive macroscopic regularities that seem to manifest a kind of group coordination or intelligence. The appeal of the analogy between humans and insects is that it helps us to see social phenomena as mere abbreviations for the complex patterns that emerge from potentially simple interactions among individuals. And this defuses anxiety about "reifying" the social world.

Nagel on Theory Reduction

If Hamilton's and Mandeville's accounts are successful, they give individualistic explanations of group properties. Ideally, the aim of such accounts was to mathematically derive the "geometry of the herd" from the behavior of individuals. This was exactly the kind of work that mid-century philosophers of science saw as the central quest of the sciences. Explanations like this formed the heart of their program of "the unity of the sciences," one of whose aims was to put dualism to rest once and for all.

The centerpiece of the program was the notion of *theory reduction*. Ernest Nagel articulated the classic model of reduction in his 1961 book *The Structure of Science*.[10] Reduction is a relation between two scientific theories. A theory, in Nagel's view, is a set of sentences expressing a set of causal laws, both experimental and theoretical.[11] A theory about schools of herring, for instance, might consist of causal laws about the behavior of the school under attack from predators:

1. When the school is in configuration A, and a predator attacks from direction B at speed C, the school will pack into a ball.

---

[9] To be precise, Mandeville does not quite present what I have called an "aggregation reversal." In his essay, it is a subset of the population being dissolute that triggers effects in the rest of the population.

[10] Nagel 1961. For a more detailed exposition, see Suppe 1977.

[11] See Nagel 1961, 79–105.

2. When the school is in configuration D, and a predator attacks from direction E at speed F, the school will form a vacuole around the predator as it swims through the school.
Etc.

A theory of individual herring, in contrast, might consist of laws about the behavior of individuals when they encounter disturbances in their immediate environment:

1. When a herring perceives such-and-such a threat, and is in the presence of some other object, it hides behind that object.
Etc.

To reduce the herring-school theory to the individual-herring theory, we need a set of "bridge laws," which satisfy two conditions. First, they connect the vocabulary of the first theory to that of the second theory. For every term of the herring-school theory (e.g., ball, vacuole, etc.), a corresponding term is defined in the individual-herring theory. For example, the term 'ball' in the school-theory corresponds to a term in the individual-herring theory, defined as a set of configurations of individual herring. The new term is defined so that whenever the individuals are in one of those configurations, the school is in a ball. Second, if we combine the bridge laws with the laws of the individual-herring theory, it has to be possible to derive from them all the laws of the herring-school theory.

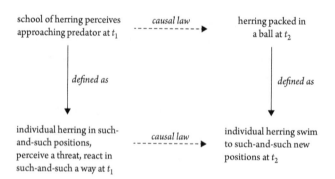

Figure 2A The reduction of "macro" laws to "micro" laws

Hamilton's geometry of the herd can be construed as a basic instance of this procedure. He begins with a simple macroscopic generalization about the packing of herds under threat. At the individual level, he proposes a mechanism for the interactions of members of the herd. Statements of these mechanisms compose the individual-level theory.

Then he gives a mathematical characterization, in individualistic terms, of the conditions for a collection of individuals to be "tightly packed." This is the definition of the new individual-level term, which corresponds to a herring-school term, for use in the bridge law. With these theories, Hamilton can derive macroscopic generalizations from the bridge law together with the individual-level theory.

Figure 2A illustrates the relations between the various parts of the theories. In the figure, the horizontal dimension represents time: the dotted arrows connect causes at time t_1 to their effects at time t_2. Usually in such diagrams, the vertical dimension is ontological: the "macroscopic" facts above stand in some ontological relation to the "microscopic" facts below. That is, they are made of them, or determined by them. This structure will show up many times later on. But in figure 2A, the relation between the macro and the micro is not exactly ontological—instead, it is linguistic: macroscopic *terms* are *defined* using microscopic ones.

Since 1961, philosophers have developed devastating objections to Nagel's approach to theory reduction. The most influential objections to Nagel's reductionism came from Thomas Kuhn, who argued in *The Structure of Scientific Revolutions* that different theories reflect different paradigms, are not objective, and are often untranslatable into one another.[12] Others challenged the mid-century views that theories involve distinct theoretical and experimental laws, that they consist of laws at all, and that they are linguistic things. Still others focused on the multiple ways that science constructs explanations, denying the centrality of reduction to the project of science. And others remained committed to theory reduction, but objected to the mid-century approach on technical grounds.[13]

These reactions have led many people to believe that the entire reductive program was a disaster, a dark chapter in philosophy. This assessment is too bleak. Within a few years, the program gave rise to innovations that have reverberated through the sciences, and social sciences in particular. It is one of those encouraging cases in philosophy where a failed program led to a much deeper understanding of the issues.

Putnam's Insight

One of the key insights to come out of the program was a new response to anxieties about dualism. As I mentioned, theory reduction gives an answer to dualism. Schools of fish do not need to be treated as separate objects

---

[12] Kuhn 1962.

[13] Garfinkel 1981; Nickles 1973; Schaffner 1967; Suppe 1972; Suppes 1967; van Fraassen 1972, 1977, 1987.

existing in a separate and real sphere, over and above the individual fish. Instead, any statement about schools of fish can be reduced to statements about individual fish. Thus reductionism gives a satisfying answer to the dualist challenge.

In the early 1960s, Hilary Putnam—who earlier had been one of the most forceful advocates of "the unity of the sciences"—began to realize two things.[14] First, that theory reduction was actually exceedingly difficult to carry out, and rarely if ever done. It was particularly not so easy to construct "bridge laws." And second, that reduction was not really needed to respond to anxieties about dualism.

One of Nagel's many accomplishments in *The Structure of Science* was to clarify the several disparate elements that must come together in a theory reduction. He noticed that bridge laws needed to satisfy a number of conditions in order for reduction to be successful. Among them was the condition he called "connectability": that every term in the high-level theory T1 must correspond to its own definition in the reducing theory T2. Nagel also pointed out that the connections between terms given in the bridge laws could have a variety of strengths.[15]

Putnam realized, however, that even though connectability might be necessary for reduction, it was more than we need if all we want is to deny dualism. Even if reduction is impossible, we can still open a window to avoid dualism by denying connectability.[16] After all, connectability is a tall order: it insists that *every term* in the high-level theory be defined in terms of the lower. As a practical matter, this has almost never been successfully accomplished. Explicit definitions of scientific terms are devilishly hard to produce.[17] But we do not need such definitions in order to avoid dualism. Even if we cannot have bridge laws between theories T1 and T2 that satisfy "connectability"—and so cannot reduce T1 to T2—we still might have reason to think that the laws described by T1 *as a whole* are taken care of by T2 *as a whole*.

Putnam argued that in many sciences, we should not even expect connectability to hold between theories at the high level and theories at the lower level. Sometimes, Hamilton-style reductive explanations might be possible, but often they are not. This is not because the objects or phenomena at the high level are metaphysically distinct, over and above the low-level stuff. It is only that the terms of the theory at the high level cannot be *defined* in terms of the

---

[14] Putnam 1967, 1969.

[15] See Nagel 1961, 353–4.

[16] Davidson 1980 presents a different influential argument for denying reducibility.

[17] Even the ones that appear somewhat straightforward are not, such as the reduction of thermodynamics to statistical mechanics. Cf. Callender 1999.

low-level theory. What might get in the way of these definitions? The fact that the high-level properties are "multiply realizable."

Multiple Realizability

Putnam's idea—which even today is probably the most widely endorsed argument against reduction—was that certain high-level properties might be "multiply realizable" at the lower level. A screwdriver, for instance, is a thing that turns screws. It might, however, be made of all different kinds of physical materials. It might be made of steel, or wood, or hard plastic, and still function to turn screws. Suppose we have a theory about screwdrivers. For instance, a theory about how different-sized flathead screwdrivers function to turn different kinds of screws. What is the physical theory that this screwdriver theory corresponds to? A theory about the properties of steel? Of wood? Of plastic? There is no single correspondence between the term 'screwdriver' and any term in any one physical theory.

Jerry Fodor filled out this argument in more detail.[18] In contrast to the neat figure 2A above, Fodor drew the following diagram (figure 2B), representing the messy relations between laws at different levels. There is not a simple correspondence between "high-level" properties and "low-level" ones. Rather, high-level properties are often realized as hugely complicated and messy properties at the lower level.

The bottom part of figure 2B is a wildly disjunctive set of causal connections. It is not a low-level causal law. Because there is no single connection between

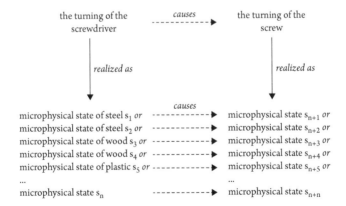

Figure 2B Multiple realizability

---

[18] Fodor 1974.

high-level and microphysical terms, the connectability condition is not satisfied. Thus the multiple realizability of properties at the high level blocks Nagel-style reduction.

Still, none of this means that we need to postulate any separate and distinct "realm" of entities at the high level. Because high-level properties are multiply realizable, reduction fails. Nonetheless, any given high-level state is realized in *some* low-level state. Any given screwdriver is made out of some particular material, not "screwdriver substance." Any given mental state is made out of some particular state of the brain, not "thinking substance." And any given social state is made out of some particular state of the population, not "social substance." Even if we cannot construct a theory reduction, we can still reject dualism.

This Putnam-Fodor argument is controversial. It has been challenged on several fronts. Many people deny that most high-level properties can be analyzed in terms of functions such as *serving to turn screws*. This may undermine the Putnam-Fodor argument for the multiple realizability of these properties.[19] Also, the notion of multiple realizability has itself come under serious criticism in recent years.[20] And, finally, a number of theorists have taken innovative approaches to reduction, so that it does not require connectability.[21] So it might be that the Putnam-Fodor argument does not actually rule out reduction, even though it rules out Nagel-style reduction.

In this complex back-and-forth, however, the basic point should not be lost. We can avoid dualism, or end up with an innocuous version of it, whether or not macroscopic theories can be reduced to microscopic ones.[22] This is a key legacy of the mid-century program in the unity of the sciences.

The Putnam-Fodor arguments dramatically clarify the point Steven Lukes made in "Methodological Individualism Reconsidered." Lukes distinguished explanatory individualism from ontological individualism, but did little more than point out that the two were distinct. Now we have a *reason* why certain reductive explanations may be impossible, even if there is nothing to the high-level phenomena over and above the low-level ones. It becomes clear how we might insert the wedge between explanatory individualism and ontological individualism.

---

[19] Theories of functions and functional properties have also changed radically since Putnam's day. See Ariew, Cummins, and Perlman 2002.
[20] See, for instance, Shapiro 2000.
[21] Hooker 1981; Schaffner 1993; Weber 2004.
[22] This result is still controversial. See, for instance, Kim 1989.

Antidualism and Supervenience

By the mid-1970s, a strategy for avoiding an unpalatable dualism was taking shape. Avoiding dualism does not require that we correlate every high-level property with a low-level property. We do not need to be able to *translate* social theories into theories about individual people. Instead, all we need is that the social properties are *collectively* "nothing over and above" the properties of individuals.

The next fifteen years saw the development of a powerful new tool for making this "over and above" phrase much more precise: the notion of *supervenience*. Supervenience is a relation between two sets of properties. Take property set A to be all the social properties and property set B to be all the individualistic properties. To say A supervenes on B, then, is to say an object cannot change its A-properties without there being some accompanying change in its B-properties. Or to put it more intuitively, *the B-properties fix the A-properties*. Once all the individualistic properties are in place, that fixes the social properties.[23]

To illustrate, consider the pictures on a TV screen. These supervene on the pattern of illumination of its pixels. The picture on the TV screen cannot change without some change in the pattern of illumination of its pixels. There is nothing to the TV picture over and above the illumination of individual pixels. Notice, however, that there need be no lining-up of individual A-properties with individual B-properties. We might have a TV image—say, an image of a ball or a tree—generated by many different patterns of illumination of pixels, on many different types of screens.

There is not just one supervenience relation, but a family of them. One member of the family is "local supervenience."[24] Local supervenience makes a claim about each object in the world. To say that A locally supervenes on B is to say that for *each object*, if you fix its B-properties, then you have fixed that object's A-properties. More relevant for our purposes will be "global supervenience." This is a weaker claim. It does not insist that the B-properties of each object fix that object's A-properties. Instead, it makes a claim only about the spread of properties across the entire world. To say that A globally supervenes on B is to say that if you fix *all the B-properties in the entire world*, then you have fixed the A-properties in the world.[25]

---

[23] See Kim 1984, 1987. Also, McLaughlin and Bennett 2005 is an excellent overview of varieties of supervenience.

[24] This is often called 'individual supervenience', but that term would be confusing in the present context.

[25] It has long been recognized that social properties fail to locally supervene on individualistic properties. This was first pointed out by Currie 1984. Though his conclusion is correct, however,

Supervenience turns out to be a hugely useful family of diagnostic tools, like X-rays or MRIs, for evaluating the relationships between "macro" and "micro" properties. As a diagnostic tool, supervenience helps show how we may be able to dodge the threat of dualism.[26]

It may be impossible to reduce social theory to a theory of individual people. But all it may take to avoid dualism is for a weaker relation to hold—a supervenience relation—between the whole set of individualistic properties and the whole set of social properties. This is the strategy endorsed by "nonreductive individualists."

To address the problem of dualism in the social sciences, the supervenience of the social on the individualistic quickly became the consensus response. It gives theorists a way to be ontological individualists, even if they are skeptical about explanatory individualism.

Contemporary Ontological Individualism

Much of this work, on reduction, multiple realizability, and supervenience, was done in the philosophy of mind. But that work has dovetailed nicely with discussions of methodological individualism in the social sciences. For instance, it is standard nowadays to argue against explanatory individualism in social science on the grounds that social properties are multiply realizable.

In recent years ontological individualism has come to be interpreted as a supervenience thesis: the social properties globally supervene on the properties of individual people.[27] The individualistic properties exhaustively determine the social properties, even if there is no way of producing a correspondence between a given social property and one or more individualistic properties. A representative statement is Harold Kincaid's:

> Social wholes are both composed of individuals and determined by their actions... Individuals determine the social world in the intuitive

---

his argument does not quite work. For a rigorous argument, see the appendix to Epstein 2011. In chapter 8, I give a more precise formulation of global supervenience.

[26] Supervenience is not a flawless diagnostic tool, but can offer excellent evidence. A crucial shortcoming is that it does not suffice for full grounding. (I discuss grounding in chapter 5, and the limitations of supervenience in chapter 8.)

[27] Without giving an exhaustive list, among those explicitly advocating global supervenience of social properties on individualistic properties are Bhargava 1992; Currie 1984; Hoover 1995, 2001a, 2001b, 2009; Kincaid 1986, 1997, 1998; List and Spiekermann 2013; Little 1991; Macdonald and Pettit 1981; Mellor 1982; Pettit 1993, 2003; Sawyer 2002, 2005; Schmitt 2003; Stalnaker 1996; Tuomela 1989.

sense that once all the relevant facts (expressed in the preferred individualist vocabulary) about individuals are set, then so too are all the facts about social entities, events, etc. Or, to put this idea in terms of supervenient properties, the social supervenes on the individual in the sense that any two social domains exactly alike in terms of the individuals and individual relations composing them would share the same social properties.[28]

Like Lukes, most people use some statement of ontological individualism as a quick prelude, on the way to discussing the obstacles to explanatory individualism. Sometimes it is accompanied by a reminder of Mandeville's point that aggregates can have emergent properties that none of the individuals do.[29]

This is the consensus view. Ontological individualism is distinct from explanatory individualism. Explanatory individualism is debatable. Some problems might be susceptible to individualistic explanation, while others are probably not. Ontological individualism, on the other hand, is much more straightforward. It is best understood as a supervenience thesis, and it is obviously true.

To be sure, ontological individualism was never intended to solve the central problems of social theory. It does not give us any indication of how to model the individual. It does not solve the problems of aggregation. It does not even suggest how to approach the longstanding debates between structure-centered and agent-centered explanations in the social sciences. The aim of the division between explanatory and ontological individualism was to free theorists from the anxieties of dualism: to clear the ring, so that more sensible fights over explanatory methodology could begin.

---

[28] Kincaid 1986, 1998
[29] E.g., Pettit 2003, 191.

3

Seeds of Doubt

Despite the appearance of its being settled, there are reasons to worry about ontological individualism. Social theorists often assume that, with a little thought, the wrinkles will be easy enough to iron out. But this attitude is too cavalier. Where theorists do make specific claims about the composition of social entities out of individuals, they tend to go wrong. Consider, for instance, Lukes's "truisms," his "banal propositions about the world that are analytically true":

> Society consists of people. Groups consist of people. Institutions consist of people plus rules and roles. Rules are followed (or alternatively not followed) by people and roles are filled by people. Also there are traditions, customs, ideologies, kinship systems, languages: these are ways people act, think, and talk.[1]

Banal they may be, but are they analytic? Are they truistic? Are they even true?

Or consider the quotation from Kincaid in the previous chapter. It is fairly clear and explicit, about as detailed as any statement of ontological individualism. But on closer look, it is vague. It says, for instance, not a thing about what gets included in "individuals and individual relations," "how things are with and between individuals," or "all the relevant facts (expressed in the preferred individualist vocabulary) about individuals." In the literature on the social world, almost no one talks about *why* social facts are supposed to supervene on individualistic ones, or even about what facts or properties count as individualistic in the first place.

In this chapter, I begin to argue what may seem like a radical claim: The contemporary consensus is mistaken. Ontological individualism is false. The social facts do not supervene on the individualistic ones. My aim in this

---

[1] Lukes 1968, 120.

particular chapter is to show this intuitively. Denying ontological individualism does not mean endorsing "ontological holism." It does not mean endorsing "emergentism."[2] What is wrong with ontological individualism is that it is a stronger thesis than many people have realized.

I will start by discussing the failure of an analogous thesis in a different science. It is an actual historical episode, one that is interesting in its own right: the nineteenth-century "cell theory" of organisms. Then I will apply that case back to the social sciences.

Why Be Skeptical about Ontological Individualism? An Analogy

As students, we all learned the principles of cell theory: *All living things are made up of cells. All cells come from preexisting cells by division. The cell is the structural and functional unit of all living things.* Cell theory is one of the great accomplishments of modern science, and was surprisingly hard won. Although Robert Hooke discovered and named cells in the 1650s, it took two hundred years for people to recognize their biological importance, and to develop these simple principles.

As responsible as anyone for this accomplishment was Rudolf Virchow, a German biologist, anthropologist, doctor, and politician. Virchow was prolific in the way that only nineteenth-century Prussians could be: a bibliography of his works runs 113 pages, his masterworks being the handbooks on pathology from the 1850s and 1860s. In *Cellular Pathology* of 1860, he cataloged the systems of the human body, presenting the cellular anatomy of each and describing the diseases and degenerations of the systems in terms of cellular transformations. He began the book with a clear statement of the primacy of cells as the building blocks of both plant and animal life.

> Every more highly developed organism, whether vegetable or animal, must be regarded as a progressive total, made up of larger or smaller numbers of similar or dissimilar cells. Just as a tree constitutes a mass arranged in a definite manner, in which, in every single part, in the leaves as in the root, in the trunk as in the blossom, cells are discovered to be the ultimate elements, so it is with the forms of animal life.

---

[2] Some commentators have misunderstood this, e.g. Hindriks 2013, p. 432; Guala and Steel 2011, 282.

Every animal presents itself as a sum of vital unities, every one of which manifests all the characteristics of life.[3]

At the time, Virchow was engaged in a struggle against theories that now seem so old-fashioned it is hard to imagine them ever being taken seriously. Although scientists had mostly abandoned ancient theories of life, such as vitalism and humor theories, cell theory was not the only alternative. The physicalist school, for instance, regarded organisms as having "formative forces." Others devalued the importance of the cell, regarding the protoplasm as the basic substance of organisms.[4] Obviously, cell theory won this struggle. But in the wake of its success, we can miss the fact that the above passage from Virchow is not quite true. It sounds plausible. But if we actually take a look at organisms, whether possums or people, it is clear that they are not a sum of cells, not even mostly.

Any organism includes a lot of extracellular material. An average human body has fifteen liters of fluid, floating and pumping around in various places. There is interstitial fluid between cells, blood plasma, gastrointestinal fluid, cerebrospinal fluid, ocular fluid, joint fluid, and urine. Metabolites, ions, proteins, neurotransmitters, and hormones flow between cells. And then there are big chunks of solid anatomical stuff that are not made up of cells either. Bone matrix—something that plays a rather important function for humans—makes up about 15 percent of body weight, but is not composed of cells. Nor are teeth. Nor are eyes—which are mostly made up of a transparent gel. Even organs like the lungs are largely built of connective tissue, consisting of fibrous proteins and collagenous compounds.

Imagine an excellent simulation, on some futuristic computer, of all and only the cells in a human body, leaving out everything that was not a cell. Red and white blood cells would be represented, but without plasma for them to travel in. There would be neurons, but no way for them to communicate with one another. There would be muscles, but nothing to connect to and pull on. There would be no chewing, no seeing, no digesting or excreting. An excellent simulation of cells would be a terrible simulation of the body.

Virchow was, of course, aware of this problem. He noticed that many tissues were made up of more extracellular material than cells. In fact, he noticed that certain tissues had very few cells at all.[5] Nonetheless, he was committed to the principles of cell theory. How was he to deal with extracellular material?

---

[3] Virchow 1860, 13–14. Emphasis in original.

[4] Drysdale 1874; Fletcher 1837. See Mayr 1982 for a fascinating discussion of views at the time.

[5] E.g., connective tissue and "mucous" tissue (Virchow 1860, 41–8).

Theodor Schwann, a few years earlier, had proposed a solution to this very problem, but Virchow found it unacceptable.[6] Schwann had suggested that the extracellular material in an organism was the stuff of *protocells*, the "cytoblastoma destined for the development of other cells."[7] It was clear to Virchow that this could not be right. Much extracellular material was destined for nothing of the sort.

So Virchow took a different strategy. Although it is true that extracellular material lies outside of cells, Virchow argued that all the extracellular material in an organism is apportioned to the cells that govern it. Cells are proprietors of "cell-territories":

> I have, by means of a series of pathological observations, arrived at the conclusion that the intercellular substance is dependent in a certain definite manner upon the cells, and that it is necessary to draw boundaries in it also, so that certain districts belong to one cell, and certain others to another . . . Any given district of intercellular substance is ruled over by the cell, which lies in the middle of it and exercises influence upon the neighboring parts.[8]

In other words, the matter between cells is divided up by boundaries, according to the cells into whose districts they fall. Virchow seems to have in mind something like the division of Lake Superior into the portion belonging to the United States and the portion belonging to Canada. Just as the two countries have "superintendence" over their respective portions of the lake, cells rule over their portion of extracellular material.

Virchow holds that extracellular material has implicit boundaries, according to the cells that have purview over it. Organisms do not only consist of the interiors of cells, but of the interiors together with the exterior parts that belong to them.[9] Thus the principles of cell theory remain unsullied: organisms are exhaustively composed of cells conceived of as cell-territories.

All in all, it was a good idea for Virchow to be as rigorously committed to cell theory as he was. He emphasized that we should see the body as composed of innumerable vital parts. Virchow's erroneous ontology was much better than the earlier erroneous ontology. But this "cell-territory" strategy is a stretch. Even if we were to believe Virchow about the governance of some extracellular

---

[6] It is an interesting question, why these biologists were so committed to cells *exhaustively* composing the body.
[7] Schwann (1839) 1847, 168; Virchow 1860.
[8] Virchow 1860, 15–16.
[9] Virchow 1860, 15–16.

material by cells in some cases, it could not apply in general. Which are the cells governing the gelatinous goo in the eye? Which cells govern the bone matrix, or the digestive fluids sloshing in the stomach? (Analogously, we might ask what countries have "superintendence" over the middle of the Pacific Ocean.)

To be sure, cells are important parts of the human body, maybe even the most interesting and dynamic parts. Certainly it would be a mistake to devalue their role, as some of Virchow's nineteenth-century opponents did. But the body is only partially composed of cells. There are many basic functional components apart from cellular ones.

Virchow's claim is not unlike a botanist claiming that trees are composed of leaves. (In fact this claim, bizarre though it seems, was defended at length by Goethe in *The Metamorphosis of Plants*.[10]) No doubt, leaves are an important part of trees. Without leaves, most trees would be in trouble. But they are only part of the story. Botany is not just leaf-ology. Nor is anatomy just cytology.

Notice that the failure of anatomy to be exhaustively determined by cytology does not involve any remarkable claims about anatomical "spirits," dualism, or ghostly ectoplasm. We would not expect anatomy to be exhaustively determined by dermatology or nephrology. There is more to the body than the skin and the kidneys. The source of supervenience failure is that the "cellular facts" are too limited to exhaustively determine the "anatomical facts."

How Bad Ontology Leads to Bad Science

Virchow's flawed ontology damaged his scientific practice. Because he insisted on the exhaustive determination of anatomical facts by cellular facts, he committed himself to thinking of the body as divided into cell-territories. And this led him onto some radically mistaken tracks when it came to certain theories.

Take, for instance, his theories of the formation and degeneration of bone. In Lecture XVIII of *Cellular Pathology*, Virchow proposes a theory of the formation of bone out of cartilage. Cartilage, according to Virchow, consists of cartilage-corpuscles, each of which is a "territory," consisting of a cartilage cell plus the "capsule" in which it is contained.[11] A bone-corpuscle, then, is formed when a cartilage-corpuscle becomes calcified. Although the cell is transformed, the territory remains the same.

---

[10] "It came to me in a flash that in the organ of the plant which we are accustomed to call the leaf lies the true Proteus who can hide or reveal himself in all vegetal forms. From first to last, the plant is nothing but leaf, which is so inseparable from the future germ that one cannot think of one without the other." (Goethe (1816) 1962, 258–9)

[11] Virchow 1860, pp. 415–18

> The old limits of the capsule still represent the real district which is under the sway of the bone-corpuscle . . . Within these limits we see the bone-corpuscle accomplish its peculiar destinies.[12]

We see the reverse process occurring, according to Virchow, in bone necroses. In degenerative osteitis, for instance, bone-corpuscles transform back into other sorts of cells.

> The bone first produced and proceeding from cartilage may undergo a transformation into marrow, then into granulation-tissue, and finally into nearly pure pus.[13]

As scientists discovered in the intervening century, there are many ways bones can degenerate. But Virchow's proposal describes none of them. When bone deteriorates, it does not change from corpuscle-sized regions of bone into marrow and other types of cell. His view of bone growth is likewise mistaken. There are, of course, cells involved in bone growth: osteoblasts act as "construction workers," depositing layers of bone. Nonetheless, a bone is no more made up of osteoblasts than a roof is made up of roofers.

Virchow's commitment to his cell-territory version of cell theory derailed his explanations. To explain bone necroses as transformations of cell territories into other cell territories is not just awkward or psychologically misleading, but is something between a distortion and a flat mistake. Ontology has ramifications, and ontological mistakes lead to scientific mistakes. Commitments about the nature of the entities in a science—how they are composed, the entities on which they ontologically depend—are woven into the models of the science. The errors of Virchow's cytocentric approach to anatomy are easy to see, as are the scientific failures that resulted. Despite Virchow's expertise with a microscope, his commitment to cell theory led him to subdivide tissues into cells where there are none. And that led to poor theories about how anatomical features come to be, how they are changed or destroyed, and what they do.

Bringing It Back to Social Theory

Virchow's approaches, strategies, and failures have close parallels in the contemporary social sciences. Take, for instance, the implicit ontology built into

---

[12] Virchow 1860, pp. 417–18
[13] Virchow 1860, p. 422

one of the best-known contemporary frameworks for social theory: the "boat diagram" developed by James Coleman.

In *Foundations of Social Theory*, Coleman observes that social theory typically intends to explain phenomena at the macroscopic level. He takes as a paradigm Max Weber's argument that the development of capitalism in the Western world can largely be explained by the influence of Protestant religious doctrine. Coleman breaks Weber's argument into three steps:

1. Protestant religious doctrine generates certain values in its adherents.
2. Individuals with certain values (referred to in proposition 1) adopt certain kinds of orientations to economic behavior. (The central orientations to economic behavior are characterized by Weber as antitraditionalism and duty to one's calling.)
3. Certain orientations to economic behavior (referred to in proposition 2) on the part of individuals help bring about capitalist economic organization in a society.[14]

Coleman points out that in this argument Weber is explaining the transition of society from one "social-level" feature to another "social-level" feature, that is, the transition from Protestantism to Capitalism. But to explain this transition, Weber moves to the "individualistic level." Weber's narrative is not a good explanation, Coleman suggests, if he does not show how the transition was mediated by the activities of individuals. This means that Weber needs to demonstrate all three propositions: how the doctrine generates the relevant values at the individual level, how the individual-to-individual transition of values leads to a certain economic behavior, and how the individual economic behavior generates capitalism at the level of the society. Coleman depicts this with the following boat-shaped diagram (figure 3A):

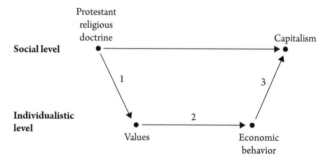

Figure 3A The boat diagram of Weber's Protestant Ethic

---

[14] Coleman 1990, 8.

The two nodes at the top of the diagram are macroscopic, social phenomena. The two at the bottom of the diagram are phenomena involving individual people. The numbered arrows roughly correspond to the three numbered propositions.

Coleman thinks Weber does a good job with arrow 1, explaining how Protestant religious doctrine leads individuals to have certain values, and with arrow 2, explaining how those individuals' values induce them to behave in characteristic ways in the economic sphere. But he complains that Weber pays inadequate attention to arrow 3. Having paid such careful attention to how Protestant values lead individuals to save and invest, Weber neglects the problem of aggregating the economic behavior of individuals into a capitalist system. This, Coleman stresses, is often the crucial problem for social theory: explaining how the systematic behavior of individuals generates systemic macro phenomena.

Coleman's scheme for social explanation is more modest than Watkins's individualistic strategy discussed in chapter 1.[15] Entities at the social level need not be eliminated, or analyzed in individualistic terms. Coleman starts his social explanations with a background of institutions that are already in place. Only the transitions marked by the arrows need to go through individuals.

Applying Boat Diagrams to Cell Theory

Nonetheless, Coleman's approach presupposes a quite specific ontology. To see this, consider what happens if we apply Coleman's diagram to the kinds of cases Virchow addressed. The first case will be a phenomenon for which cell theory is adequate, and the second, a phenomenon for which it is not.

For the first, consider the organism-level phenomenon: *Being deprived of water causes a plant to wilt*. We might construct an explanation along the lines of figure 3B. There is an easy transition from the organism-level cause, *being deprived of water*, to the cellular-level effect, that cells cannot replace the water they lose. Arrow 2 represents causes that take place entirely at the cellular level, where water loss causes shriveling. Arrow 3 represents the plant-level effect of shriveling at the cellular level. This is the sort of explanation that Virchow might have successfully produced.

---

[15] Following Karl Popper, Coleman endorses a local or situational approach to explanation rather than insisting on reductionism. See Agassi 1975; Jarvie 1998; Popper (1957) 2002.

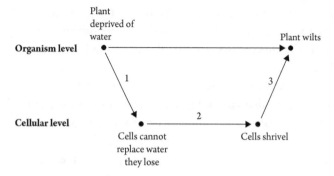

Figure 3B A successful boat-style explanation at the cellular level

Contrast this, however, with the sort of case that trips up Virchow's version of cell theory (figure 3C). Consider the organism-level phenomenon: *Being deprived of fluoride causes a child to get cavities.*

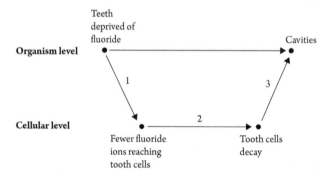

Figure 3C A failed boat-style explanation at the cellular level

In this diagram, we again employ a causal transition from the organism level to the cellular level, one within the cellular level, and one from the cellular level back up to the organism level.

But this diagram is absurd. The problem, of course, is that tooth enamel is not made of cells. So it is not only this explanation, involving fluoride ions reaching tooth cells, that is spurious. It is unlikely that there will be *any* explanation at the cellular level for tooth decay. The diagram implies that what is predominantly causally important for tooth decay is cellular. This is something Virchow might have believed, but we know better.

Now, the advocate of the boat diagram might object that I am being too literal. Here are some natural reactions one could have:

1. *When we speak of the "cellular level" in an explanation of tooth decay, we mean anything cell-sized. Even though there aren't tooth cells, there are cell-sized structures, and we can explain tooth decay in terms of those structures.*
2. *When we speak of the "cellular level," we don't really mean the **cellular** level. We mean the microscopic level, the level of whatever parts the tooth is made up of. The point, after all, is that there is some lower level explanation for tooth decay, and that explanations in lower level terms are good ones.*
3. *When we speak of the "cellular level," we mean not just cells, but anything that interacts with cells. And there are cells involved in tooth decay—epithelial cells, gum cells, nerve cells, cells of the bacteria the form plaque, and so on. All of those are at the cellular level, and so tooth enamel is as well.*

These are all variants on a theme. The first reaction is particularly reminiscent of Virchow's cell-territory response. It did not work when we said that explanations had to go through cells, so we expand what we mean by 'cell'. But the other reactions are similar. They all ask us to take the word 'cellular' with a grain of salt.

To a limited extent, these reactions make sense. We do not want to be so uptight as to rule out pragmatic explanatory strategies like Coleman's boat just because we do not like their labels. Still, it is fair to ask, if we can be as flexible as we like in what we include in the "cellular" level, then in what sense is the explanation an explanation in terms of *cells*? We can twist words, but the facts are the facts: when it comes to tooth decay, not much happens at the cellular level. What is interesting is happening at large-scale levels, such as pits forming in tooth enamel, and at very small-scale levels, such as at the level of interactions between enamel and acids. But teeth are not composed of cells. This means that figure 3C is misapplied, not that we should start playing games with what counts as "cellular."

To endorse a boat-style explanation is to endorse an ontology. If a good explanation of high-level phenomena has to go through the lower level, that presupposes that facts at the lower level exhaustively determine those at the higher level.[16]

What goes for cellular boat diagrams goes for individualistic boat diagrams as well. Coleman praises Weber's overall strategy in the *Protestant Ethic*, faulting it only for its failure to explain inference 3. Implicit in this, however, is a presupposition: the phenomena at the social level are more like wilting plants than like decaying teeth. But if social phenomena are not exhaustively determined by facts about individuals, then we should not expect an explanation that goes by way of

---

[16] It actually assumes much more: that the entire causal network that relevantly affects the system is also exhaustively determined by the lower level of entities.

individuals to work. His explanatory strategy carries with it a commitment to a particular ontology of the social world. If that ontology does turn out to be mistaken, then the explanatory strategy will have to be revised or abandoned.

An Intuitive Failure of Individualism

Coleman's explanatory strategy carries ontological presuppositions. *If* social facts are not exhaustively determined by facts about individuals, it does not make sense to insist that social explanations should conform to the boat diagram. But why think that the claim *social facts are exhaustively determined by facts about individuals* is analogous to the erroneous claim *anatomical facts are exhaustively determined by facts about cells*?

Here is a reason: consider facts about the Starbucks Corporation. On a typical day at Starbucks, pots of coffee are being brewed, baristas are preparing frappuccinos, cash registers are ringing, customers are lining up, credit cards are being processed, banks are being debited and credited, accountants are tallying up expenses, ownership stakes are changing in value, and so on. At least on the face of it, some of these facts about Starbucks fail to supervene on facts about the people and their interrelations. To be sure, the employees are critical to the operation of Starbucks. But facts about Starbucks seem also to depend on facts about the coffee, the espresso machines, the business license, and the accounting ledgers.

Consider what we might want to accomplish in a model of some changing property of Starbucks. Suppose we were to model it through some sort of unfortunate event. Suppose, for instance, there is a freak, late-night power spike at a number of Starbucks outlets, causing the blenders and refrigerators to break, the ice to melt, and the milk to spoil. Suppose that event is the last straw for a financially struggling Starbucks, underinsured as it is. So, when the power spikes and its key assets melt down, its assets no longer exceed its liabilities. Overnight, as the owners, employees, and accountants are asleep in their beds, Starbucks goes from being financially solvent to insolvent.

In this example, the transition to insolvency involves property and equipment, not individuals. It is analogous to the tooth-decay argument in this respect. At least at first blush, it is not individuals, or phenomena at the "individualistic level," that explain this social-level transition. If this is right, a Coleman-style explanation for the transition in terms of social facts would be impotent (figure 3D).

Like the explanation of tooth decay, the explanation of Starbucks's transition from one macroscopic state to another does not make sense, if given at an

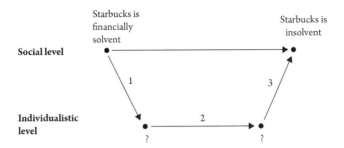

Figure 3D A failed individualistic explanation of Starbucks

irrelevant or incomplete microscopic level. It is not, of course, that no people at all are involved in the ordinary course of Starbucks operations. Nor is it that there are no cells involved in the ordinary course of anatomical operations. But just as that fact does not entail that anatomical explanation should go through the cellular level, neither does the involvement of people in Starbucks entail that explanations of its states should go through the individualistic level.

Also analogous to the anatomical case is the point about simulation. Imagine an excellent simulation, on the same futuristic computer, of all and only the people involved in Starbucks—employees, customers, board members, even vendors—but leaving out everything that is not a person. It might be organized something like figure 3E.

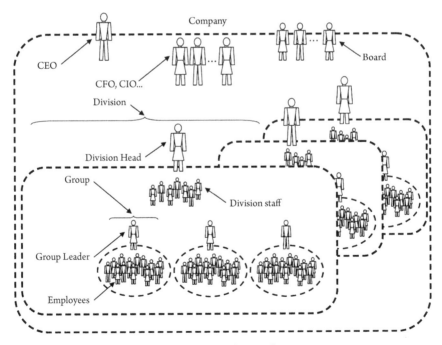

Figure 3E Multiscale organization with employee subagents. From: Michael J. North and Charles M. Macal, *Managing Business Complexity* (New York: Oxford University Press, 2007)

In this simulation, there would be baristas, but no coffee to drink. Customers, but no chairs or wi-fi connections. Cashiers, but no cash. However excellent this might be as a simulation of the people, it would be a terrible simulation of the company.

And also as in the tooth decay case, there are several Virchow-style reactions one might have. This is a common move in social theory: we bend what is meant by "individualistic." *When we speak of the "individualistic level," we don't really mean the* **individualistic** *level. We mean individual-sized things, whatever they are. Or we mean the microscopic level, the level of whatever parts society is made up of. Or we mean anything that causally interacts with individuals.* In short, just as Virchow slides from cells to cell-territories, we slide from individuals to individual-territories, or else from individuals to things that are not in any respect individualistic.

Such are the moves social theorists have implicitly taken. Even Watkins did this, in the later part of his career. Back when he was debating Mandelbaum in the 1950s, he argued that social facts are entirely composed of the psychological states of individual people. Over time, however, he realized that psychological states were an inadequate base for determining facts about the social world. After all, social phenomena involve behaviors and actions, not just thoughts. It is absurd to think that social facts—like actions performed by Starbucks Corporation—are nothing more than the psychological states of individual people. At the very least, even the most strident individualist needs to admit non-psychological stuff, like bodily movements. Quietly, Watkins started to expand his notion of what counts as individualistic beyond the psychological.[17]

But his moves were not enough. In the intervening years, epicycle upon epicycle has expanded, in one way or another, the set of individualistic facts. Some theorists emphasize that when a customer purchases a drink from a barista, the two people interact with one another; they do not just stand isolated from one another. Thus, they argue, we should take the individualistic facts to include certain relations between and among individuals.[18] Others point out that social facts depend on the parts of the world we interact with. In microeconomics, for instance, we typically model economies not just as interacting individuals, but also include "bundles of resources" they own. Here the individual-territories include not just people, but the "bundles" they govern.

Other theorists point out that we need a more sophisticated theory of beliefs and other mental attitudes, beyond simple psychological states. Philip Pettit, for instance, has developed a theory he calls "individualistic holism," where

---

[17] E.g., in Watkins 1959.

[18] This move accomplishes less than is usually thought. See my discussion of Hodgson in Epstein 2014a.

the mental attitudes of individuals are themselves social.[19] Still other theorists take the basic building block to be at once individualistic and yet somewhat broader than the bodies and minds of individual people. Anthony Giddens, for instance, builds social facts out of "practices," which are the patterns of activity of individuals in the world.[20]

All of these theories take the whole of the social world to be carved into individual people and—in one way or another—their respective "territories." Yet it is unclear why any of these should succeed.

Admittedly, some social phenomena *do* seem to be naturally divisible into individual people, or into individuals and their resources. For instance, a flea market is a bunch of tables piled with goods, each table manned by a vendor who owns the goods being sold. Individual customers walk around the tables, holding some money, and sometimes exchanging that money for goods.

But many things in the world are unlike flea markets. Starbucks outlets, for instance. Or air force battles: these do not naturally break down into individual people. People are involved, but the basic units of action seem to be planes and aircraft carriers, not people. It seems more natural to see the battle as pieces of military hardware interacting with one another, with the people acting as resources apportioned to them, than the interaction of people. Or economies: many economists take these to be divisible into individuals, households, firms, and institutions, each with its own bundle of resources. Yet it is not obvious why this should be so, any more than for Starbucks.

This much is only an intuitive point, a seed of doubt about ontological individualism. Some theorists will regard the comparison with Virchow to be slander. Others will defend Virchow and the comparison. Still others will wonder whether there is not a different sense in which the social world is made from individual minds—namely, that it is a projection of our minds onto the natural world. (This is an issue I take up in the next chapter.) But in all this, one thing is clear: we cannot trust the prevailing dogma. We cannot trust it—not until we engage in more careful metaphysics.

A more careful metaphysics is best done from scratch. It is a waste of time to haggle endlessly over the meaning of 'individualism', or to trace the paths of the dozens of individualistic theories in circulation. Instead we ought to just cut through it. Given that ontological individualism is a claim in metaphysics, we might as well avail ourselves of the latest technology in metaphysics, rather than shy away from it. We should even add to that technology when needed.

---

[19] Pettit 1993.
[20] Giddens 1984.

4

Another Puzzle: A Competing Consensus

The prevailing dogma cannot be trusted. Ontological individualism was a major advance over those old theories that mixed up claims about ontology with claims about explanation. And its enduring and widespread popularity should make us wary of a quick dismissal. But this does not mean it is true. Nor does the absurdity of social dualism make ontological individualism true. Ontological individualism needs to be examined suspiciously and closely, precisely because it has *individualism* at its core. So, it seems, we have a long road ahead of us.

At this point, however, I want to take a right turn. In the last three chapters, I have been speaking about *the* consensus, *the* settled view, of the nature of the social world. That was not quite accurate. Ontological individualism is almost universally endorsed. But it is not the only consensus claim about the social world. A second view is also almost universally endorsed. The philosopher Francesco Guala, in fact, has gone so far as to call this second view the "Standard Model of Social Ontology."[1]

The idea is this: the social world is a kind of projection of our thoughts, or attitudes, onto the world. We, as a community, *make* the social world by thinking of it in a particular way. The bills in my pocket are money because we all think of them as money. The president has the powers he does because we grant him those powers. America is a nation because we think of it as such. The social world, quite generally, is the social world in virtue of our beliefs about it.

Strangely, we rarely notice that this thesis differs sharply from ontological individualism. To many people, the Standard Model seems like a version of ontological individualism, a particular way to fill out the details. But it is not. It is a different claim about a different aspect of social metaphysics. Moreover, the Standard Model offers a different response to dualism about the social world.

[1] Guala 2007.

Just like ontological individualism, it denies an autonomous or separate sphere of the social. But the Standard Model's denial of dualism exploits a different strategy than does ontological individualism. In fact, the Standard Model is at odds with ontological individualism, even though both views are endorsed by many of the same people. Ontological individualism does not logically contradict the Standard Model, but if one is right, it is very likely that the other is wrong.

Instead of one consensus view, we have two. And they are in tension with one another. With this, we find a big monkey wrench in the works. Theorists largely see themselves as agreeing on the basics of social ontology. But they do not have a consistent picture of what they are agreeing on.

There is, however, good news. By sorting out the conflict between these perspectives, we can cut the Gordian knot. We can quickly assemble a synthesis, a model of the social world with several parts that work together. Ultimately, the Standard Model will fall, just as ontological individualism will. But by seeing how they address complementary problems, we can pave a shortcut for clarifying both.

In this chapter, I present this Standard Model of Social Ontology, and explain how it differs from ontological individualism. I do so using two versions of the model: John Searle's and, going back a few centuries, David Hume's.

Searle's Version of the "Standard Model"

In *The Construction of Social Reality*, Searle proposes a theory of what he calls "institutional facts."[2] This is a broad category that includes many of the things in the social world. Universities, governments, restaurants, and money are all examples of institutional facts. Searle contrasts institutional facts with "brute facts." A dollar bill is an institutional fact, and a piece of paper with green printing on it is a brute fact.

According to Searle, members of a community create institutional facts in their community, by imposing "statuses" on material objects. His simplest example is a boundary line around a village. Searle tells the following story. In ancient times, the inhabitants of some village built a high wall to keep out invaders. It worked because the wall was high. The wall physically functioned to keep the invaders out. Over time, however, the wall deteriorated. Eventually, it was only a line of stones in the sand surrounding the village. But the villagers and their neighbors had grown accustomed to having the wall there. Despite the fact that the line of stones no longer physically functioned to keep people out, the villagers and their neighbors continued to treat the line of

---

[2] Searle 1995. He gives a slightly revised theory in Searle 2010.

stones as a boundary, just as if it were the wall. At this point, the line of stones had taken on a symbolic function. Even though the line of stones provided no physical barrier to movement, the villagers and their neighbors have imposed the functional "status" that it once had as a physical object: the status of being a boundary.[3]

This is an intriguing story, and with it Searle introduces the centerpiece of his theory: the *constitutive rule*. The constitutive rule expresses the status that the villagers impose on the physical object, that is, on the line of stones in the sand. The constitutive rule for boundaries, according to Searle, is: "*The line of stones **counts as** having the status of functioning as a barrier **in** the village.*" Saying that X counts as Y in C basically means that, in context C, we treat X as if it performed the function we associate with Y, and hence we give it a certain status. The "X term" denotes the object to which the status is assigned. The "Y term" denotes the status assigned to it. We have institutional facts because we have constitutive rules in place in our society.

From this toy example, Searle moves to a more realistic one: paper money. In creating money, according to Searle, we assign a status to pieces of paper that have been printed in a certain process. Dollar bills, for instance, are just pieces of paper issued by the Bureau of Engraving and Printing. But when a piece of paper is issued in this way, we assign it a very important status. Dollar bills, according to Searle, have the following constitutive rule:

(CR) Bills issued by the Bureau of Engraving and Printing (X) count as dollars (Y) in the United States (C).[4]

The form of the constitutive rule, however, is only part of the theory. The constitutive rule expresses what object or objects receive what status. But what is it that puts constitutive rule (CR) in place, in a community? What makes (CR) a constitutive rule for dollars? This is the second part of Searle's theory.

Constitutive rules are not facts of nature. Instead, Searle argues, what puts a constitutive rule in place in a community is that we *collectively accept* it. That is, constitutive rule (CR) is in place in the United States because the people in the United States all have a particular attitude toward (CR). "Something is money," Searle explains, "only because we think of it as money."

---

[3] Searle 1995, 39.
[4] Searle 1995, 28. It is questionable whether Searle gets the conditions for being a dollar bill right, but (CR) is fine for illustrating his view.

The other centerpiece of Searle's theory, then, is an account of "collective acceptance."[5] Searle argues that for a community to "collectively accept" something is not for just a bunch of individuals to have the attitude "I accept the rule." In Searle's view, for the members of a community to collectively accept something is for each of the community members to have a special kind of attitude toward it: a "we accept" attitude.[6]

To summarize, Searle's theory consists of two parts. One part is about *constitutive rules*, which have the form *X counts as Y in C*. The second part is about what puts the constitutive rule in place in our community. What puts a constitutive rule in place, according to Searle, is that we *collectively accept* that rule. Collective acceptance is the glue that binds a constitutive rule to a community.

Another Version of the Standard Model: Hume's Theory of Convention

Searle's is not the only version of the Standard Model of Social Ontology. To broaden our view a bit, it is helpful to look back to an older theory, historically even more influential than Searle's. In book III of *A Treatise on Human Nature*, David Hume presents a theory of government, justice, money, property, promises, and languages. Hume argues that these things are created by *social conventions*.

Consider, for instance, his theory of promises. Hume begins with a distinction: a person's promise to do something, he points out, is different from a mere resolution to do something. If I promise to paint your house, that involves my resolution to paint your house, but it also involves more. If I resolve to paint your house, I am not *obliged* to paint it. Whereas if I promise to paint your house, I have incurred an obligation to paint it.

According to Hume, social convention is what adds obligation to promises. On Hume's account, we have a social convention of the following form: words uttered according to a certain formula incur an obligation. That is, when somebody utters a phrase of the form 'I promise to S', that utterance is a promise.[7]

Before Hume, political theorists had applied the notion of "convention" to certain kinds of laws, and occasionally to language as well. Hume expands it

---

[5] Searle modifies this slightly in Searle 2010.

[6] This analysis of collective attitudes is controversial. Still, nearly everyone agrees that collective acceptance of a rule is not just a matter of each person having an individual-acceptance attitude toward it. I discuss collective attitudes in chapters 14–16.

[7] See Hume (1740) 1978, book III, part II, section V.

to a much broader array of social phenomena. Property, money, justice, and so on, are all products of social convention.

What is a social convention? It is here that Hume makes the biggest contribution. Historically, theorists had thought of convention as a matter of agreement—either explicit or tacit. Hume, however, severs the connection between convention and agreement. A convention, for Hume, is instituted by some regular behavior, together with beliefs on the part of members of the community that the behavior is in their mutual interest. The recognition by each member of the community that performing the activity will be mutually beneficial, and the expectation that other members of the community will perform the activity as well, provides reason to perform the activity. A convention, says Hume in the *Enquiries*, is:

> a sense of common interest; which sense each man feels in his own breast, which he remarks in his fellows, and which carries him, in concurrence with others into a general plan or system of actions, which tends to public utility.[8]

For a convention to be in place, in Hume's view, is thus a matter of certain beliefs being held by members of the community, accompanied by regular behaviors.[9] In the case of promises, the regularities and beliefs put in place the convention that when somebody utters a phrase of the form 'I promise to S', that utterance is a promise.

So Hume's theory can be divided into two parts, just as I described in connection with Searle's. First, as with Searle's constitutive rule, Hume has a principle or rule that is established: *When somebody utters a phrase of the form 'I promise to S', that utterance is a promise.* Second are the facts about our beliefs and actions that put the convention in place. In this case, *English speakers believe that it is mutually beneficial to take utterers of the formula 'I promise to S' to incur obligations to perform S, and they behave accordingly.*

Hume's account gives us the basics of a theory that at once illustrates the Standard Model, and at the same time departs from Searle's theory. In Searle's view, the constitutive rule is entirely put in place by the possession of certain "acceptance" attitudes by members of the community, toward the constitutive rule. Hume's theory differs in two notable ways. First, the attitudes are simpler on an individual level: each member has a belief and an expectation about the

---

[8] Hume (1777) 1975, 257.

[9] Hume does not explicitly say that those beliefs can be tacit, but makes it clear that he thinks they can. (See Hume (1740) 1978, book III, part 3, section 2.) Moreover, tacit conventions were discussed long before Hume; see, for instance, Pufendorf (1673) 2007.

behavior, but does not have to have an attitude directed toward the rule. The rule emerges from a matrix of common beliefs, without the rule itself having to be the object of attitudes. Second, Hume's theory of what puts a convention in place involves more than attitudes. In Searle's theory, constitutive rules are put in place *only* by attitudes. Hume, on the other hand, takes conventions not to involve just our attitudes, but also material facts about regularities in our practices.

Summarizing the Standard Model

In an insightful discussion of the topic, Francesco Guala highlights three characteristics of this Standard Model of Social Ontology.[10] First, according to this model, the social world is made by our attitudes. In Searle's view, for instance, we have collective attitudes about what makes something money, and those attitudes are the very things that put the constitutive rule in place.

Second, the model holds that the social world is *performative*: "If social entities are made of beliefs, they (unlike natural entities) must be constantly re-created (or 'performed') by the individuals who belong to a given social group."[11] One point here is that social things can be created by linguistic performances or declarations. The official utterance "I declare you man and wife" does not just describe a marriage, but creates it. Another point is that the social world must be actively maintained. In Hume's theory, a convention is sustained just as long as members of the community continue to have the appropriate beliefs and practices regarding it.[12] Searle has a similar view. In order for a constitutive rule to be maintained in our community, we need to continue to accept that rule. That continued acceptance does not need to be at the forefront of our minds.[13] We can accept things tacitly, as well as explicitly. But without continued acceptance, the constitutive rule is no longer in place in the community.

Third, the model takes the social world to be the product of *collective intentionality*. This is the more general topic, and both Searle's theory of collective acceptance and Hume's theory of shared belief are instances of it. All the theories in the Standard Model hold that we need to have some collective attitudes or intentions in order to constitute the social world.

---

[10] Guala 2007, 961–3.
[11] Guala 2007, 962.
[12] Hume (1740) 1978, book III, part 3, section 2.
[13] Searle 1995, 117–19, 127–47.

Taking all these together, the Standard Model provides a particular picture of the nature of the social world: the social world is made and maintained by us, by our mental attitudes. And not just our attitudes as individuals, but as a community.

The Tension between the Two Consensus Views

Many things are appealing about this Standard Model. But how does it relate to the theories of the social world in the last three chapters? The way this model approaches the making of the social world is distinct from the way ontological individualism does. Ontological individualism is about one relation between individuals and the social world, and the Standard Model is about an entirely different relation. Consider the following sets of examples:

1. a. A *mob* of drunken hockey fans storming down Howe Street in Vancouver, breaking windows and overturning cars
   b. The *flow of commuters* in the Boston metropolitan area, moving in and out of trains, subways, and buses
   c. The *Jewish people*, expelled from Spain in 1492 and migrating into Europe and North Africa
2. a. A *handicapped parking spot*, marked by a blue and white sign
   b. A *tea party* arranged by a group of children, with stuffed animals arranged around a table, each with a miniature tea cup and saucer
   c. An *unkosher* animal, such as a lobster or pig

All of these involve social objects in some sense "made" by people: *a mob, a flow of commuters, a people, a parking spot, a tea party, an unkosher animal.* But the two sets of examples involve people in different ways. The things in Set 1 are composed of people. The mob is composed of drunken fans, the flow of Boston commuters is composed of travelers, and the Jewish people is composed of various people with a certain religious/ethnic background. The things in Set 2, on the other hand, are *not* composed of people. The parking spot is a section of pavement, with some paint marks. The tea party is a bunch of stuffed animals and cups. (The animals, not the children, were sitting at "tea.") The unkosher animal is a real animal.

The sets also differ in how they are conceptualized, or "borne in mind." The parking spot is authorized by the city, and marked out by city workers. The tea party is orchestrated by the children, who tacitly agree on their game. The laws of kosher food are set out in Leviticus. The drunken mob, on the other hand, is more like the school of selfish herring. Although each individual is intelligent (or somewhat intelligent) few if any conceptualize the mob as a whole. The same holds true for the commuters and the Jewish people. These people may or may not conceive of themselves as a people or group.

These are two very different relations between the people and the social phenomenon they "make." In a very loose and casual sense, we might say that the fans "constitute" the mob, and the children "constitute" the tea party. But these turn on two different meanings of 'constitute'. Consider how we might depict the relation between facts in the two situations. Type 1 examples can be depicted quite simply (figure 4A).

Figure 4A Depicting Type 1 examples

This pair of facts is related in the way that ontological individualists, like Watkins, Lukes, and the supervenience theorists, were suggesting. The properties of the mob, like *running down Howe Street*, supervene on the properties of Bob, Jane, and the others. There is nothing to the social facts over and above the facts about the individual people.

Type 2 examples, however, cannot be depicted so simply. Here is a first pass at drawing the relations between a fact about the tea party, and the nonsocial facts that "make" that social fact (figure 4B).

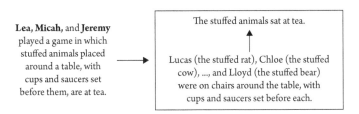

Figure 4B Depicting Type 2 examples

In the second diagram, the roles of people are different from their roles in the first. (I have highlighted the people in bold letters, to make it clear where they are.) In the first diagram, there is nothing to the mob "over and above" Bob, Jane, and the others. In the second diagram, we can see two different relations at work. The upward-pointing arrow does the same thing as in the previous diagram: it connects the facts about the things that "compose" the tea party to the fact about the tea party. The tea party is not composed of people. The rightward-pointing arrow represents a different relation. Lea, Micah, and Jeremy are not the constituents of the tea party, but their game is the "glue" that holds together what is going on in the box. Lea, Micah, and Jeremy's mental attitudes set up the game, put in place the "constitutive rule" for what it takes to be at tea. But the things that satisfy the X term of that constitutive rule are not mental attitudes. They are stuffed animals.

Ontological individualism holds that social facts supervene on facts about individual people. The property *being a mob* supervenes on properties of individual people clustering together. The same cannot be said for *being a tea party* or *being a dollar*, according to the Standard Model. It does not supervene on mental attitudes alone.

What does *being a dollar* supervene on? Among the things it supervenes on are the properties of pieces of paper. Given that the constitutive rule for dollars is what it is, for something to be a dollar requires that it be printed in green ink, on a particular kind of paper. Moreover, it supervenes in part on the properties of the Bureau of Engraving and Printing.

According to Searle's version of the Standard Model, the fact that the community collectively accepts rule (CR) puts in place that rule. That is, it sets up the conditions for something to be a dollar. Even after the rule has been collectively accepted, however, we do not yet have dollars. To get those, we need the X part of the constitutive rule to be satisfied. We need green pieces of paper, and issuance actions by the Bureau of Engraving and Printing.

The Standard Model, in other words, does not say that social facts *are* collective intentions. It says that collective intentions *set up* conventions, or constitutive rules. These conventions or constitutive rules are propositions about what sorts of nonmental, nonpersonal stuff constitute social things. This, in short, is the tension between the two different consensus views. Ontological individualism holds that social facts supervene on facts about individuals. The Standard Model holds that facts about individuals set up the conditions for something to count as a social fact. They offer two distinct approaches to the social world.

Another conflict is in how the respective views respond to dualism. Ontological individualism addresses dualism using the strategy described in chapter 2: social facts supervene on individualistic facts. The Standard Model also addresses dualism, but differently. It takes the social world to be a projection onto the nonsocial facts. On the Standard Model, the social world is "brute" facts treated in a certain way.

That does not mean that Standard Model theorists take the social world to be a fiction or an illusion. Two things are meant to prevent that. First, social facts are the product of collective, not just individual, attitudes. (On the Standard Model, if I alone accept that Bob is president, then that is a fiction, but somehow, if we as a group accept that Bob is president, then that is a social fact.) Second (and more persuasively), on the Standard Model the social world is not an attitude in our heads, but the actual stuff in the world to which a certain status or convention has been assigned. Take away the line of stones, and there is no boundary. The boundary is not an illusion; it is just the line of stones taken against the background of our constitutive rules or conventions.

Where We Stand

If the Standard Model is so different from ontological individualism, why have they been conflated with one another? The main reason, in my view, is that they have both remained so loose and unclear. In social theory, we have been cavalier about the metaphysics and not taken the details seriously enough. To be sure, there are other reasons as well. Both models appear to be responses to dualism. Both are in a sense "individualistic." Both seem to apply to certain widely discussed cases. But for all these similarities, the views are different enough that if we talked about them precisely, their differences would pop out.

A lack of clarity is endemic in all sides of the literature. Consider, for instance, Searle's term 'institutional fact'. Examples of institutional facts, according to Searle, are money, marriages, restaurants, nations, national boundaries, and so on. This is very confusing terminology. Money is not a fact—it is a social object, or maybe a social kind. *I have a dollar in my pocket*: that is a fact, a social fact. *A dollar exists*: that is a different social fact. *The bill in my pocket constitutes a dollar*: yet a different social fact. But *dollar*: that is a thing or a kind of thing, not a fact. And *being a dollar*: that is a property. If we want to inquire seriously about the making of social facts, objects, or properties, we need to be clear about what we talking about.

Even the best discussions in the literature are loose about the details. Consider, for instance, the passage from Harold Kincaid I quoted above. Although it seems to be written very precisely, it really is quite confusing. Here are three important ways it is unclear:

1. *It is unclear about what it means for something to **determine** something else.* Kincaid says, "Social wholes are composed of individuals and determined by their actions." What does this sentence mean? If "social wholes," whatever they are, are composed by individuals, then what about them is "determined by their actions"? Does Kincaid mean that they are *caused* to be what they are by actions? Or does he mean that they are somehow *composed* by actions? Yet if they are composed by actions, then why did he say that they are composed of individuals? Then in the next sentence he says that they

are *determined* by individuals. Why the flip between composed and determined, and between individuals and actions?
2. *The role of* **language** *is unclear.* Kincaid speaks of facts being expressed "in the relevant individualist vocabulary." What does vocabulary have to do with it? If some fact is the case, what does it matter whether it is expressed in one vocabulary or another? For instance, if the fact *I am six feet tall* is "set," then it makes no difference to the fact being set if we express it as "I am six feet tall" or as "I am 72 inches tall." Does it?
3. *It is unclear about the* **sorts** *of social and individualistic things we are talking about.* In the quoted passage, Kincaid moves from a point about "social entities, events, etc." to a point about "supervenient properties." Why the change to properties, and is the thesis about properties meant to be equivalent to the preceding thesis about social entities and events? Just what sorts of social things are determined by what sorts of individualistic things? Facts? Events? Properties?

My point is not to pick on Searle and Kincaid: their discussions of the social world are among the best out there. But if we are to move past intuition and to seriously understand society, we cannot be so loose about the building blocks we choose to work with. Most crucially, we need to avoid mixing up talk about the world and talk about language. And we need to be clear about what things we are talking about—facts, events, properties, relations, objects, and so on.

5

Tools and Terminology

This chapter offers a few basic tools of metaphysics, including the following:

- The distinction between facts, propositions, and sentences
- Possible worlds and possible facts
- Properties and relations
- Social facts and social kinds
- The grounding relation

This is a long list. Rather than giving a complete or systematic presentation, I will try to offer a reasonably precise way to make and assess claims about the nature of the social world. There are many models we might use for accomplishing this, but as much as possible I will stick to the standard interpretation of the standard tools. (The discussion of "grounding" is a minor exception, for reasons I will discuss.)

A Three-Part Model: Facts, Propositions, and Sentences

Let's start with a couple of facts: *The Earth is round*, and *Bill Clinton was president of the United States in 1994*. These are facts about the world. We use language to talk about them, but they are facts about a planet, a shape, a person, and a social property, not facts about language.

A three-part model for organizing these points distinguishes (1) the world, (2) abstract representations of the world, and (3) language, or ways of speaking about the world. According to this model, the sentence 'Bill Clinton was president of the United States in 1994' expresses a proposition. The same proposition can be expressed with other sentences as well. For example, 'The President of the United States in 1994 was Bill Clinton', or 'Bill Clinton était président des États-Unis en 1994.' Because it is a fact that Bill Clinton was president of the United States in 1994, the proposition representing that fact is true.

Facts are things in the actual world. Propositions are abstract representations of the world. Some propositions are true and some are false. The true propositions are the ones that correspond to the facts, and the false propositions are the ones that do not. Each fact corresponds to a true proposition.[1]

Sentences are expressions in particular languages, but propositions do not depend on language. Before there were people, there was no language, and no sentences to express propositions. But there were still true and false propositions.[2] The proposition expressed by the sentence 'The Earth is round' would have been true even if people had never existed, because it is an abstract representation of the fact that the Earth is round. Which would be true even without us. Some propositions are about language, like the proposition expressed by the sentence, '"A" is the first letter of the alphabet.' But many propositions are not.

Propositions are not linguistic, nor are most of them about language. Still, we do use language to express propositions. Furthermore, we also use language to *denote* facts and propositions, i.e., to specify which fact or proposition we are talking about. When I am being precise, I will use italics for propositions, and bold italics for facts. For instance:

(5.1) The proposition *The Earth is round* is true.

(5.2) The fact ***The Earth is round*** obtains.

Notice that (5.2) does not say that the fact ***The Earth is round*** is "true." Facts are not true or false, any more than a chair or a lake is true or false. An object exists and similarly a fact obtains, or is the case. A proposition corresponding to a fact is true if and only if that fact obtains.

Possibilities

Some propositions are true, and some are false. For example, the proposition *Bill Clinton was president of the United States* is true, and the proposition

---

[1] This is standard, but there are many ways of analyzing facts and propositions. See Neale 2001; Richard 1990.

[2] Many people are bothered by the idea that there could be propositions without us. Again, all this talk about propositions can just be regarded as a useful model for thinking about the world. Lots of metaphysicians have developed models that circumvent propositions, but they tend to be much more complicated than the standard ones are, and not as powerful.

Chelsea Clinton was president of the United States is false. Equally, we can distinguish (actual) facts from possible facts. Possible facts are things that might obtain in the world, some of which actually obtain and others of which do not. For example, the possible fact **Bill Clinton was president of the United States** actually obtains, whereas the possible fact **Chelsea Clinton was president of the United States** does not. But both are alike in terms of being *possible* facts.

Some people are queasy about talk of possibilities, and about treating possible facts similarly to how we treat facts. But this model is convenient. Possible facts correspond to propositions, and each possible fact that obtains in the actual world corresponds to a true proposition. This is particularly helpful for thinking systematically about the sciences. When we build models in the sciences, we are not only interested in the way things actually are, but how they *would be* if things changed. How would unemployment change if we raised interest rates? How would educational attainment change if we increased standardized testing? These are questions about ways the world might be. Similarly, we might model the circumstances in which **Chelsea Clinton is president of the United States** obtains. In such a model, we could consider what other facts have to obtain in order for it to obtain.

Another convenience is to talk about "possible worlds" as a whole, and to compare them to one another. A good deal of debate in metaphysics is dedicated to the question of how to understand possible worlds. Are they fictions? Concrete objects? Linguistic constructions? Complex properties? Shadowy "ways things might be"? For our purposes, it does not much matter which of these we adopt. We can avail ourselves of talk like "In some possible worlds, Chelsea Clinton is president of the United States," and "In all possible worlds, two plus two equals four," without committing ourselves to a particular theory of possible worlds, or possible facts.[3]

It is now standard to cash out the ideas of necessity, contingency, and impossibility, in terms of possible worlds. Whether a proposition is true or false often depends on the way the world is, although some propositions are true or false regardless of the way the world is. For instance, the proposition *Bill Clinton was president of the United States* is true. If things were different in the world, that proposition would be false. Such a proposition is *contingent*. The proposition $2 + 2 = 4$, on the other hand, is true however the world might be. This proposition is *necessary*—that is, true in every possible world. And the proposition $2 + 2 = 5$ is false however the world might be. It is, in other words, *impossible*—false in every possible world.

---

[3] This should be qualified: see the arguments against "linguistic ersatzism" in Lewis 1986, 142–65. Also see Sider 2002.

Properties and Relations

Just as the person Bill Clinton is not the same thing as the name 'Bill Clinton', we also need to distinguish properties from the predicates we use to denote them. The expression 'being president of the United States' is a predicate. It is a linguistic item. That predicate denotes a property, that is, the property *being president of the United States*.

The predicate 'is taller than' expresses a relation between two objects, not a property of a single object. The predicate 'is taller than' is called a binary (or two-place) predicate, and the relation *is taller than* is called a binary (or two-place) relation. Thus the sentence 'Bill is taller than Hillary' consists of two names and a binary predicate, and the proposition *Bill is taller than Hillary* consists of two objects—Bill and Hillary—and a binary relation that holds between them—the *is taller than* relation. Sometimes I will flank the relation with letters, to make it clear that is a binary relation: for instance, the relation *A is taller than B*.

There are also three-place predicates and relations, four-place predicates and relations, and so on up. For instance, if Carol is sitting between Bob and Alice, then the three-place *sitting between* relation holds of the three objects Carol, Bob, and Alice.

Objects actually have (or instantiate) some properties, and possibly have others. Chelsea Clinton, for instance, has the property *being the daughter of Bill and Hillary*, and possibly has the property *being president of the United States*. In any possible world, any given object either does or does not have any given property. Additionally, it should be stressed that how a property is instantiated does not change over time. Of course, the properties that a given object instantiates can change over time. For instance, the color of a particular piece of clothing may fade over time, for example from bright red to pale red to pink. But what it takes for something to have the property *being bright red* remains invariant over time.

To put this point differently, we might associate with a property a set of "instantiation conditions." For a given property P, the instantiation conditions are simply other properties, such that necessarily an object has P if and only if that object has properties R and S and T, for example.[4] Instantiation conditions like these do not change over time or possibilities. Even a property that involves a particular time, such as *being president of the United States in 1994*, has fixed instantiation conditions.

---

[4] I do not assume that all properties have such sets of associated properties, nor that we can always know when a property does.

I will use this same model of properties for social properties, just like any other property. For instance, Bill Clinton changed from having the property *being governor of Arkansas* to *being president of the United States*. But the conditions for his having either of these properties did not. "But suppose," one might object, "the Arkansas legislature changed the law about how governors are elected. In that case, wouldn't it be true that the instantiation conditions for the property *being governor of Arkansas* have changed?" Not according to the model of properties we are using. To make sense of this, we can understand the instantiation conditions to be something like *satisfying whatever conditions the Arkansas legislature sets out for being governor*. Or we can simply take there to be two different properties having different instantiation conditions, the property *being-governor-of-Arkansas-before* and the property *being-governor-of-Arkansas-after*.

This model is, of course, an idealization. We do identify the before and after versions of such properties, as being a change in a single property. But there are two good reasons for pushing that burr under the carpet. First, in order to make sense of it, we would need a more sophisticated apparatus than I can discuss in this book. We would need a model of how a given property can be "anchored" and "re-anchored," over and over again, while remaining the same property. (Anchoring is a topic I introduce in the next chapter.) That, however, would require us to revise more metaphysics than is practical or needed for now. So, for our purposes, properties have unchanging instantiation conditions, both over all times and across all possibilities.

Second, there are advantages to this model of properties—that is, as having unchanging instantiation conditions over time. When we want to assess whether a given object has a given property, we want to do it in a single way, regardless of the time or circumstances we are assessing. Otherwise there is no way to compare whether things have changed. Consider, for instance, a proposition about some change over time, something like *Clothing has become more brightly colored over the course of the last century*. We do not want the truth or falsity of that proposition to depend on changing conditions for what it takes to be brightly colored. Instead, we take the property *being brightly colored* to have fixed instantiation conditions, throughout that period.[5]

So long as we are being clear about properties and relations, we should also be clear about what sorts of things they apply to—that is, their *relata*. For instance, I have already mentioned the supervenience relation. The literature

---

[5] That does not, however, mean that when I talk about properties I mean only intrinsic properties. (On the distinction between intrinsic and extrinsic properties, see Kim 1982; Langton and Lewis 1998; Lewis 1983; Yablo 1999.) Both intrinsic and extrinsic properties have fixed instantiation conditions.

sometimes gets quite confused because it is vague about the relata of this relation, that is, what supervenes on what. As I pointed out, supervenience is best (and most commonly) understood as a relation between sets of properties. But this is often muddied, with people talking about it as relating sets of facts, events or even sets of predicates. Although these may seem innocuous, they are not, as I will discuss later on.

Social Facts

In the passage I quoted earlier, Kincaid says: "Once all the relevant facts (expressed in the preferred individualist vocabulary) about individuals are set, then so too are all the facts about social entities, events, etc." But when is a given fact a fact *about* individuals? Kincaid seems to suggest that it has something to do with being expressed in "the preferred individualist vocabulary." But this cannot be right. Facts are unaffected by the way we describe them, in the same way that the moon is unaffected by the ways we describe *it*—e.g., calling it 'la luna'. Fortunately, the model I have described points us in a more fruitful direction. Consider an example from the last chapter:

(5.3) **Bob, Jane, Tim, Joe, Linda, ... and Max ran down Howe Street.**

(5.4) **The mob ran down Howe Street.**

Are (5.3) and (5.4) the same fact? It is understandable why one might think they are. Both obtain, and at least in some limited sense, there is nothing more to the mob than those people. But there are at least two reasons that (5.3) and (5.4) denote different facts. One is that there is a possible world in which (5.3) is the case and (5.4) is not, and another possible world in which the reverse is the case. (Consider a world in which Bob and the others are running down different parts of Howe Street, maybe in different directions. Though they are all running down Howe Street, they are not a mob. And consider a world in which the mob is running, but Bob leaves the mob.) If they denoted the same fact, one could not obtain without the other.

Here, however, I also want to highlight another reason these are two different facts. In the model I have described, facts correspond to propositions, and the corresponding propositions are different. One is a proposition having Bob, Jane, and the others as constituents,[6] and the other is a proposition having the

---

[6] Here I am speaking of propositions as though they were Russellian, i.e., having objects and properties as constituents. Similar points can be made with different conceptions of propositions as well, but this is a convenient one.

mob as a constituent. There are two different propositions, and hence there are two different facts: one about individuals and one about a mob.[7] Notice that this has nothing to do with language, or the vocabulary in which anything is expressed. The way we distinguish facts from one another corresponds to the way we distinguish propositions from one another, not the way we distinguish sentences from one another.

As a rough guide, we can take a *social fact* to be a fact that corresponds to a proposition that has any social entity as a constituent. It might have social objects as constituents, or it might have social properties as constituents, or both. Of the four facts listed in table 5a, the first three are all plausibly social facts, and the fourth is not:

Table 5a **Social vs. nonsocial facts**

	Social object	Social property
The mob was impoverished	x	x
The mob was cold	x	-
John was impoverished	-	x
John was cold	-	-

Some people talk only about social objects, and some only about social properties. But if we are talking about social facts, we cannot limit ourselves to just one or the other.[8]

Social Kinds

It is common nowadays for philosophers to speak of "social kinds" or "human kinds." This is a term introduced by analogy to "natural kinds," which have long been discussed in metaphysics and philosophy of language. The intuitive idea of a *natural kind* is a category of objects grouped naturally, rather than arbitrarily or by fiat. Typical examples given of natural kinds include kinds in physics, such as electron and charge, kinds in chemistry, such as gold and water, and kinds in biology, such as the various species. A gold ring, a gold

---

[7] This presumes, of course, that the mob is not identical to the individuals. I discuss this point in chapter 10.

[8] As I discuss in chapter 8, this trips up Jaegwon Kim's treatment of fact supervenience in Kim 1984.

bar, and a gold nugget, for instance, all group together into a category because of their chemical composition, which does not depend on human choices or interests.

It is controversial which kinds are natural, and even whether there are any natural kinds at all. But, these debates notwithstanding, natural kinds seem to have some distinctive characteristics. John Stuart Mill, for instance, observed that they form the basis for inductive inferences in the sciences. By investigating certain gold things, testing and analyzing their characteristics, we can draw inductive inferences about other gold things. Another characteristic natural kinds seem to have is that they are essential to their members. If something is a piece of gold, then it is *essentially* a piece of gold: it could not be changed into lead without being destroyed. This is a more controversial thesis, and the relation between natural kinds and essentialism remains hotly debated.

The notion of a social kind is somewhat looser. It is convenient to distinguish social kinds from social properties more generally, largely because social kinds seem to figure in the social sciences similarly to how natural kinds figure in the natural sciences. Social scientists commonly use terms like 'class division', 'religiously sanctioned inequality', 'economic factor', 'material circumstances', 'public good', 'commodity', and so on. These terms are grammatically similar to natural kind terms, and the things they refer to seem to work in inductions, just as natural kinds do. On the other hand, it is certainly wrong to think of these categories as being independent of human activity, and it is not clear that they have the other distinctive characteristics that natural kinds do.

In speaking about social kinds, then, it is useful to think of them as the categories we might use in the social sciences, but remain open-minded about the sorts of categories these might be. Maybe the social kinds are the same as the social properties. Maybe they are a subset of the social properties. Or maybe they are a different thing altogether. To make progress, we do not need to start with a comprehensive understanding. We can just regard "social kind" as a generic way of referring to categories like these.

Social kinds—like social properties—have fixed instantiation conditions. Or, more appropriately, we might say that kinds have fixed *membership conditions*. The conditions under which something is a member of a social kind are the same across all times and possibilities. The reason for this is the same as the one I gave above. Social kinds serve a variety of functions: we employ them for recognizing things, classifying things in various situations, finding and correcting departures from norms, drawing inductive inferences, and accomplishing other practical matters.

As such, they are applicable across a universe of different situations: we can look at any object whatever, in any situation, and assess whether that object is a member of the kind *teacher, tire, hem,* or *hipster*. That does not mean that social kinds are not put in place by local contexts in the actual world. The kind *hipster*, for instance, is put in place by a range of idiosyncratic facts about our current society. But its potential instantiation is not limited to that current situation. The membership conditions for *hipster*, in other words, are what they are for a panoply of reasons. But the conditions are the conditions, and we can look around at all possible objects in all possibilities to see if those conditions apply, not just at the objects in our local context.

Grounding

Consider again the following two facts:

(5.3) *Bob, Jane, Tim, Joe, Linda, . . . and Max ran down Howe Street.*

(5.4) *The mob ran down Howe Street.*

As I pointed out earlier, (5.3) and (5.4) are different facts. Nevertheless, they are intimately related. On Tuesday evening, both obtained at 10:00 and were the case until 10:25. Then, at 10:26 neither obtained. Then at 11:08 they obtained again, and at 11:37 did not obtain.[9] In fact, every time those people ran together down Howe Street, the mob did. An amazing coincidence!

Of course, the coincidence is not so amazing. These are not two arbitrary facts, but are related to one another in a particular way. Fact (5.4) obtains at 10:00 p.m. *because* (5.3) does. The two facts are metaphysically related to one another. They are not the same fact, nor is (5.3) quite sufficient for (5.4) to obtain. If Bob and the others disperse, then even if (5.3) is the case, (5.4) might not be. So to be more precise, (5.4) obtains at 10:00 *in part because* (5.3) does.

We use the word 'because' in many different ways. Often, it connects causes and effects. We say that the barn burned down because Mrs. O'Leary's cow knocked over the lamp. The lamp is causally related to the fire, not metaphysically related to it. Knocking over the lamp "makes" the fire in a causal sense, not a metaphysical one. The flames, on the other hand, are metaphysically related to the fire. The flames do not cause the fire; in a sense, they *are* the fire.[10]

Facts (5.3) and (5.4) are not causally related to one another. The fact that Bob and the others ran down Howe Street was not the causal reason that the mob did, but the metaphysical reason.

---

[9] I am being a bit casual about the role of time.
[10] For more on the distinction between causal and constitutive relations, see Bennett 2011; Haslanger 1995; Koslicki 2012; Schaffer 2012.

To assign a word to this "metaphysical reason" relation, we say that fact A *grounds* fact B. The fact **The barn door, walls, and roof are burning** grounds the fact **The whole barn is burning**; the fact **I am married** grounds the fact **I am not a bachelor**; and the fact **a million herring turned in such-and-such directions** grounds the fact **the school split in two**. All of these pairs of facts are metaphysically related in the sense that the first fact "metaphysically makes" the second fact the case. Grounding will be a central part of the discussion from here on, so I will say a bit more about a few of its characteristics.

1. Fundamentality

Grounding is usually understood to involve a kind of priority, or fundamentality. The more fundamental fact grounds the less fundamental fact. For instance, the fact **A million herring turned in such-and-such directions** grounds the fact **The school of herring split in two**, but not the other way around.

This is perhaps the most controversial part of the notion of grounding, since many people worry about the idea that some parts of the world are "more fundamental" than others. In developing a model for thinking about social facts, however, it is hard to imagine how we could do without a notion of fundamentality. The fact **Bob and Jane ran down Howe Street** is part of what we look for in seeking the metaphysical explanation for the fact **The mob ran down Howe Street**. But not the other way around: the fact **The mob ran down Howe Street** does not metaphysically explain the fact that Bob and Jane did.

2. Partial vs. Full Grounding

In some of the examples of grounding I have mentioned so far, the grounding facts are metaphysically sufficient for the things they ground. In the following case, for instance, the first fact is metaphysically sufficient for the second:

(5.5) **I am married.**

(5.6) **I am not a bachelor.**

Two facts, however, can stand in a relation where one is *part of* the metaphysical explanation for the other, but is not quite metaphysically sufficient. For instance, the following pair of facts:

(5.7) **Bob and Jane ran down Howe Street.**

(5.8) **The mob ran down Howe Street.**

Fact (5.7) is part of the metaphysical explanation for (5.8), because Bob and Jane were participants in the actual event. On its own, however, (5.7) is not enough. Two people do not make a mob. In this case, we say that the former fact *partially grounds* the latter fact. (Sometimes I will talk about fact G "fully grounding" fact F. But there is no difference between *G fully grounds F* and *G grounds F*. I add the word 'fully' just to make the contrast with partial grounding clear.)

Looking at some of the other cases, it actually takes some thought as to whether they are examples of full or partial grounding. For instance, look again at (5.3) and (5.4). Suppose that (5.3) is the case, but that all the people are actually running in different parts of Howe Street—some at the end, others a mile behind them, and others at the beginning. Then, even though (5.3) obtained, (5.4) might not. Another fact should be added to (5.3), in order to fully ground (5.4). For instance, the fact that all those people were clustered together. Or else the fact that other running people filled in the gaps between Bob, Jane, Tim, and the others. On its own, (5.3) only partially grounds (5.4). Together with one of those other facts, it fully grounds (5.4).

Here I will mention a point of departure between the way I will use the grounding relation, and the way it is usually understood in the literature. Most people take full grounding to involve necessity. That is, if fact F fully grounds fact G, then it is *necessary* that if F obtains, then G obtains. In my view, this is not the best way to understand full grounding.[11] Instead, I will distinguish full grounding from an even stronger relation, the relation *A determines B*. Determination, as I will discuss later, is basically just full grounding plus necessitation. There are practical reasons for making this distinction, but it is easier to explain this in the context of examples I will present later on.

So for now, I will just flag the point. If F fully grounds G, then F is metaphysically sufficient for G. That does not mean that in every possible world where you have F, you also have G. Still, (full) grounding is a pretty strong relation, and it is most important to keep grounding distinct from partial grounding.[12]

---

[11] For a detailed discussion of the controversy, see Skiles 2014. Also related are: Audi 2011; Correia 2005; Dancy 2004; deRosset 2010; Fine 2012; Leuenberger 2014; Rosen 2010; Witmer, Butchard, and Trogdon 2005; Zangwill 2008. I favor the "contingency" rather than the "necessitarian" view, but none of the substantive claims in the book turn on it. In particular, the division between grounds and anchors (see chapters 6 and 9) is an independent point.

[12] For a fuller discussion of the intuitive notion of grounding, see Audi 2011 and Rosen 2010. For a more complete overview of the details, see Fine 2012.

3. Evidence for Grounding

How do we tell if one fact partially grounds another? How do we tell if one fact fully grounds another? There is no infallible method that works in all cases. But there are ways of working it out. One way is just to think things through. Is fact (5.3) metaphysically sufficient for (5.4)? Or is there something to (5.4) over and above (5.3)? Sometimes we can come up with conceptual justifications for various grounding claims. Another method is to apply certain tests, or diagnostic tools. One especially valuable test, for instance, is to examine the ways that various facts vary in lockstep with one another.

Above I pointed out that (5.3) and (5.4) co-vary with each other over time. By that I mean that when one obtains the other also does. And when one does not obtain, the other does not. This co-variation is decent evidence that there is some relation between the two facts. On the other hand, there can be lots of reasons for co-variance even without grounding. They might be causally connected: where there's smoke, there's fire, but the presence of smoke and the presence of fire are causally related, not metaphysically. Or two facts might co-vary by accident.

Nonetheless, different sorts of co-variation can be useful tools for diagnosing different sorts of grounding relations. When we consider the co-variation of facts, we can consider how facts change over time, but also how facts change over different possibilities. This is what supervenience is about. Supervenience is built using the idea of co-varying properties: a set A of properties *supervenes* on a set B of properties just in case any change in the A-properties must be accompanied by a change in the B-properties.[13] As such, it is a diagnostic tool for assessing whether facts of the form ***x has such-and-such an A-property*** are grounded by facts of the form ***x, y, z, . . . have such-and-such B-properties***. (More accurately, it is a test for metaphysical dependence. But dependence can also be evidence for determination, so supervenience is a useful tool for diagnosing both. I discuss these topics in more detail in chapter 8.)

Moving On

To make headway, it is crucial to work with a simple and powerful toolkit, to be precise, and to apply the tools consistently. With regard to the nature of the metaphysical tools themselves, I tend to approach them with a lightly accepting temperament. For instance, do propositions really exist? Some people

---

[13] More precisely, a change in the pattern of A-property instantiations must be accompanied by a change in the pattern of B-property instantiations.

insist they do, while others scoff. For our purposes, neither hill is worth dying on. To develop a model for making sense of the social world, I use propositions and these other tools freely, without worrying here about the commitment to a rich ontology.

It may seem ironic or even hypocritical to be casual about the "reality" of propositions, while refusing to be casual about the "reality" of governments, money, and other social entities. But this little hypocrisy is worth the payoff, at least to get us going. These tools of metaphysics are powerful for their precision, and for how much detailed work has gone into assembling them into a cohesive model. They enable us to think about the social world in much clearer ways than without them. And we have to use some toolkit or another, so we might as well start with a powerful one. We use the best tools at our disposal to investigate things they might help with. And then with those new insights, maybe over time we can scrutinize the tools themselves with more success.

Now let's return to the two models discussed in the last chapter, applying these new tools to assemble the models into a unified picture. That will allow us to discuss the anchoring relation and the notion of a frame principle.

6

Grounding and Anchoring

Before the interlude on tools and terminology, we were confronted with two different approaches to the social world: ontological individualism and the "Standard Model." I described ontological individualism in the first two chapters and raised worries about it in the third. It is a widely held thesis, its content largely agreed on: it is a thesis about supervenience. That is, there can be no change in the social properties without a change in the individualistic properties. The "Standard Model of Social Ontology," alternatively, is the view that social objects are projections of our attitudes or agreements onto the nonsocial world. Social entities, on this view, are performative and the product of collective intentions.

I suggested that these two views hinge on two different relations between people and the social world. The first view takes facts about people to be the *building blocks* of social facts. In the second view, facts about people's attitudes *set up* constitutive rules or conventions governing the social world. A mob is a paradigm of the first view, and a dollar bill is a paradigm of the second.

With more tools in hand, let us now get more specific about these relations, and how they fit together in an overall model of social facts. In the model I will put forward, all social facts involve both relations. Any given social fact has building blocks, and also metaphysical reasons for why that fact's building blocks are what they are. This chapter explains the overall model and its parts.

What "Constitutive Rules" Are After

Consider again the Standard Model. Searle's version of it, for instance, consists of two parts. There is the constitutive rule, having the form *X counts as Y in C*, and there are the facts about collective acceptance, which put the constitutive rules in place.

One of the problems I raised with Searle's view was that it was unclear about what sorts of social things it was supposed to give an account of. Searle characterizes his view as a theory of "institutional facts." But boundaries and money themselves are objects or kinds, not facts.

It turns out that facts are a good category for us to focus our attention on. But if we are talking about facts, we should talk about *facts*, not objects or kinds. As a working example, it makes sense to pick something simple—say, a fact about a particular dollar bill. Take the dollar bill in my pocket. Call it 'Billy'.[1] A nice example of a particular social fact is this: **Billy is a dollar bill**.

With this particular fact in mind, consider the constitutive rule for dollars. Remember that Searle's proposal is:

(CR)   Bills issued by the Bureau of Engraving and Printing count as dollars in the United States.

Let's not worry about whether Searle's rule gives the right conditions for a bill to be a dollar. It probably does not, since it is unclear whether that bureau is really responsible for issuing currency, or whether it is just the organization that prints currency. But let's not press this point.

Instead, let us consider what Searle is trying to capture or accomplish with the constitutive rule, and assess whether his proposed formula actually does that. First off, we should note that his idea would have been much clearer if he had written the formula as a conditional—an if-then statement—rather than as what linguists call a "generic." One issue with the "generic" formulation is that the formula *X counts as Y in C* is meant to apply both to particular cases and to general cases. It is meant to be a formula for constitutive rules applying to one particular thing, such as a particular line of stones counting as a boundary. And it is meant to be a formula for all the bills issued by the Bureau of Engraving and Printing. So it is not clear whether 'X' is a singular term or a general term. But in any case, generics are notoriously difficult to interpret. They often hold only for some of the Xs, not all of them. (For instance, the generic "Mosquitos carry malaria" is true, even though most mosquitos do not.) A universal conditional is much more explicit, something like: *For all objects z, if z has property X, then z has property Y.*

That much is just a technical preference. But when we cast it in more explicit form, we start to see what Searle is trying to do with the "constitutive rule." It is intended to give the conditions an object needs to satisfy in order for it to be an instance of the relevant social kind. That is, for an object to be a dollar bill, it needs to satisfy the antecedent conditions given in the "X term." If an object

---

[1] Here I am glossing over a subtle issue—which object exactly does the name 'Billy' refer to? The dollar bill, or the piece of paper that materially constitutes the bill? I discuss material constitution in Part Two, and a precise formulation of constitutive rules will need clarity on this.

has the property *being issued by the Bureau of Engraving and Printing*, then that suffices for it to be a dollar bill.

The point may be even clearer with Hume's example of a promise. In Hume's theory, we have the convention *If an uttered phrase has the form 'I promise to S', that utterance is a promise.* This is a statement of the conditions that something needs to satisfy in order to be a promise. If something is an uttered phrase having that form, then it is a promise.

In both of the cases, we can see that the aim of this part of the theory is to give the conditions for something to have a social property. But there is still something missing. The aim of the "constitutive rule" is not just to give a set of *happenstance* conditions for something to be a dollar bill, or a promise. Instead, it is to give the conditions for *grounding* a fact about a dollar, or about a promise. The antecedent is not just an accidentally sufficient condition. It is the metaphysical *reason* that something is a dollar, or a promise. Constitutive rules tell us what grounds what.

Grounding is most straightforwardly understood as a relation between facts. And in investigating social metaphysics, we look for the reasons for a wide variety of social facts to be the case. This is what a constitutive rule should be telling us. Sometimes we set up grounding conditions for a *particular* fact. For instance, we set up grounding conditions for the existence of one particular boundary around a village. More typically, we set up general conditions for grounding some *kind* of social fact. If we consider particular facts like **Billy is a dollar**, or else **Joey is a dollar**, these both obtain because an object satisfies the appropriate grounding condition. The fact **Billy is a dollar** is grounded by the fact **Billy was issued by the Bureau of Engraving and Printing**. The fact **Joey is a dollar** is grounded by the fact **Joey was issued by the Bureau of Engraving and Printing**. But both of these fall under one constitutive rule.

In short, a typical constitutive rule articulates the link between a set of grounding conditions X and a grounded fact of type Y (figure 6A).

For any z, the fact z is X grounds the fact z is Y.
Grounding conditions   Grounded fact

Figure 6A Parts of a constitutive rule

A constitutive rule is a principle that connects a set of grounding conditions to a particular social fact or a type of social fact. It articulates what the grounding conditions are for a social fact. This means that the constitutive rule is not among the grounding conditions for a social fact,[2] but instead describes

---

[2] I explain and defend this in detail in chapter 9.

how the social fact is grounded. Moreover, it expresses the grounding conditions across an entire set of situations, contexts, or worlds. We might depict the grounding of a range of different social facts with the following diagram (figure 6B).

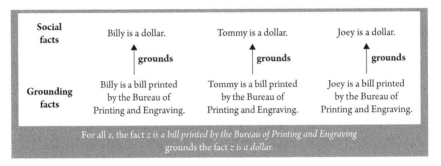

Figure 6B Grounding facts about dollars, in a context or world

This figure depicts several facts about dollars being grounded. These are actual facts. Billy is the bill in my pocket, Tommy is the bill in the drawer, and Joey is the bill on the table. All three were issued by the Bureau of Engraving and Printing, hence all three are dollars. In this context, the various social facts are all grounded in accordance with a single principle. But the principle is not "in the picture," alongside the particular instances of grounding. Rather, it is part of the "picture frame." It is a principle that expresses the grounding conditions that have been set up.

In talking about these principles, I am shifting away from Searle's term 'constitutive rule' altogether. As we will see, there are many different sorts of principles that give the grounding conditions for social facts. Many of them serve quite different purposes from the ones Searle discusses. The term 'constitutive rule' is so closely associated with Searle's formula *X counts as Y in C*, that it would be very confusing to retain his term for this much more general notion. Also, the term 'constitutive rule' was never a particularly appropriate one. Like the Holy Roman Empire, which was neither holy, nor Roman, nor an empire, constitutive rules are neither constitutive nor are they rules. Instead, I will call these general principles "frame principles."

Frames and Frame Principles

Figure 6B displays only the grounding of *actual* facts. But the grounding conditions for ***x is a dollar*** do not only apply to actual dollars. They also apply to other possibilities. The Bureau of Engraving and Printing, for instance, might have issued more dollars than it actually did. Or it might have issued fewer. These kinds of possibilities are exactly the sorts of things that we want to model in the social sciences. If we want to work out the consequences of different policies, we consider what would happen in other possibilities—for instance, what chain of events a different dollar-printing policy would unleash. Is it a good policy choice to fire up the printing presses, and crank out the bills? In examining that possibility, we apply the same frame principle to a nonactual situation.

More generally, social kinds can be instantiated across the universe of different situations, contexts, or worlds. When we set up the conditions for some social fact to obtain, we set up the grounding conditions for that universe. We set up the conditions for it to obtain even in situations, contexts, or worlds where we do not exist.

This means we have to generalize from a single situation, context, or world to a universe of possible worlds. A *frame* is a structure containing this universe, that is, a set of possible worlds in which the grounding conditions for social facts are fixed in a particular way.[3] Each of these possible worlds may have different grounding facts from one another. (For instance, in a different possible world, it is not the piece of paper Billy that was printed by the Bureau of Engraving and Printing, but instead a different piece of paper, Mary.) Different possible worlds, therefore, may have a variety of different social facts that are thereby grounded. Across the entire frame, however, the grounding conditions for social facts are the same.[4]

A frame principle gives the grounding conditions not just for the actual world, but for all possibilities. Figure 6C depicts an entire frame, all governed by one frame principle.

In this diagram, several different possible worlds are depicted. In different worlds, different grounding facts obtain. In some worlds, both Billy and Tommy are issued by the Bureau of Engraving and Printing. In some worlds, different objects are. These different facts about bill issuance ground different facts about dollars. The same frame principle, however, governs all these possibilities. The kind *dollar* has one fixed set of grounding conditions, which can be satisfied by different facts in different possible worlds. These grounding conditions are given by one frame principle.

This much—frames, frame principles, grounding conditions—is already the start of a framework for thinking about social metaphysics in general. It immediately raises a number of topics to investigate. What are the grounding

---

[3] Frames can be modeled using a multiframe or multidimensional modal logic. For background on multimodal logics, see Blackburn, van Benthem, and Wolter 2006; Marx and Venema 1997; van Benthem 1996. For an application to social ontology, see Grossi 2007; Grossi, Meyer, and Dignum 2006.

[4] If the "contingency" view of grounding is correct, this need not always be so. But for our purposes, I will take all the frame principles to be necessary in the frame.

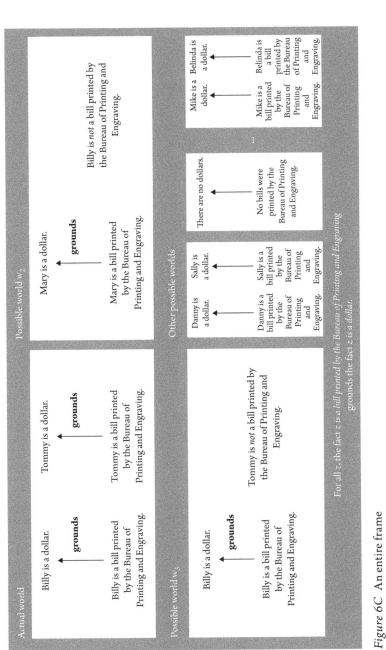

Figure 6C An entire frame

conditions for a given social fact? What *kinds* of social facts are there, and how do we express the various grounding conditions for them? However, I should note that we are far from finished with refining the form of frame principles. Several issues need to be addressed in order for the form to make sense. But we will be able to make quicker headway on these in the context of the specific case of groups, which is the topic of Part Two of the book. Instead, I now turn to the other part of the model—why are *these* the grounding conditions for *x is a dollar*, rather than *those*? What puts frame principles in place?

Anchoring

Recall the diagram we drew for the children's tea party in chapter 4 (figure 6D).

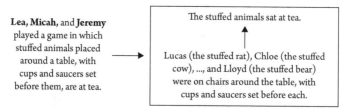

Figure 6D The role of people in the "tea party"

In this figure, we can now discern a grounding relation on the right hand side. In the box on the right, two facts are listed:

(6.3) **Lucas, Chloe, . . ., and Lloyd were on chairs around the table, with cups and saucers set before each.**

(6.4) **The stuffed animals sat at tea.**

The children have set up a game in which facts about the arrangement of the stuffed animals ground facts about their being "at tea." In setting up the game, Lea, Micah, and Jeremy have set up not only the parameters for the *actual* moves the animals make. They set up the grounding conditions for different possibilities. For example: *If the plastic triangle were placed in front of the stuffed rat instead of in front of the stuffed bear, then the rat would be "eating a scone" and the bear would be "eating nothing."* The right-hand box could therefore be elaborated with a diagram analogous to figure 6B, depicting the rules of the game as the frame principles, operative across the whole range of possibilities. None of this, however, yet includes Lea, Micah, and Jeremy. They are not part of the facts that ground the "tea party" facts. Rather, they *set up* the rules: the frame principles. It is because of facts about the children that the frame principles are in place. The rightward arrow in figure 6D represents a different relation than grounding.

It is a general feature of kinds—not just social kinds like dollars and play tea parties—that something needs to *glue* them together. Even a natural kind like gold may need a bit of "glue," to set it up as a natural kind. Some philosophers hold, for instance, that laws of nature play some role in acting as this glue. The idea is that all it takes for an object to be a sample of gold is to be composed of atoms with a particular atomic number. However, what unifies a chemical kind (like gold) into a natural kind is that the laws of nature make the chemical behave in certain regular ways. Without laws gluing the chemical kind together, it would not be a natural kind at all.[5]

The Standard Model gives us a standard answer about how the social kinds are "glued." They are glued by our ongoing attitudes toward those kinds. Searle and Hume give us different versions of this theory. Recall that in Searle's theory, they are glued together by a very particular fact: the fact *The members of the community collectively accept the constitutive rule for the kind*. On Hume's theory, they are glued together by the beliefs and practices that put in place a convention about that kind. There are surely other possibilities as well. But all of these theories are theories about a particular relation. They are theories about the "putting in place" relation that holds between a set of facts and the grounding conditions for a kind—in other words, between a set of facts and a frame principle. This is the relation I call *anchoring*.

'Anchoring'—like 'grounding', 'causing', and many other terms in metaphysics—is difficult to define explicitly. The way we fix the reference of a term in metaphysics is not very different from the way we fix a term in the sciences. We do it by describing it and pointing to it. Philosophers have not yet, for instance, worked out an adequate analysis of grounding. Some of its basic characteristics are still unknown. Does the grounding relation hold only between a more fundamental fact and a less fundamental fact? Or is fundamentality irrelevant to grounding? These questions are still being investigated, but that does not imply that they do not have answers. The term 'causation' is similar. We have scads of examples of events that stand in causal relations to one another, but there is basic disagreement about the characteristics of causation. Still, that does not prevent the term 'causation' from picking out a particular metaphysical relation, even though we do not quite know what it is. Just as we do not need to have a perfect theory of temperature, lightning, or magnetism, in order to refer to and theorize about them, neither do we need a perfect theory in order to start theorizing about causation, grounding, and anchoring. (Of course we cannot need a perfect theory of something in order to start theorizing about it. If we did, we would never be able to start theorizing about anything.) Instead, we pick these things out with examples, partial theories, and metaphors.

---

[5] This is only one view among many of natural kinds.

In the case of anchoring, theories stretch back at least to Locke's account of nominal essences, if not back to Aristotle's agreement-based theories of language and of law. But the distinction has not been clearly made between the grounding facts, and the facts that put in place the grounding conditions.

It is natural to wonder whether there really is a difference between anchors and grounds. Why aren't anchors just more grounds for a given social fact? In the examples I have given, I have tried to present an intuitive case for this. But the question remains a good one: I will address it head on in chapter 9. At the end of the day, however, what matters is whether the model works. If anchoring (or grounding, for that matter) turns out to be very useful, that provides evidence that we are in the vicinity of something illuminating. So it is as important to apply it, and see how it works in particular cases, as it is to argue for it abstractly.

I will take anchoring to be a relation between a set of facts and a frame principle.[6] For a set of facts to anchor a frame principle is for those facts to be the metaphysical reason that the frame principle is the case. In this sense, anchoring is very much like grounding. For a set of facts $g_1, \ldots, g_m$ to ground fact f is for $g_1, \ldots, g_m$ to be the metaphysical reason that f obtains in a world. For a set of facts $a_1, \ldots, a_n$ to anchor a frame principle R is for $a_1, \ldots, a_n$ to be the metaphysical reason that R holds for the frame. Both are "metaphysical reason" relations. But they do different work, and stand between different sorts of relata.

Putting the Picture Together

Anchoring and grounding fit together into a single model of social ontology. To illustrate, let us use Searle's theory of dollars as an example once again. According to that theory, the fact **We collectively accept CR** anchors constitutive rule CR. That constitutive rule expresses the grounding conditions for facts about dollars. It is then particular facts about pieces of paper (i.e., **these bills were issued by the Bureau of Engraving and Printing**) that ground the social facts (i.e., **these bills are dollars**). Depicting this involves adding just one element to the diagram: the anchor. The frame principle does not stand alone, without metaphysical explanation. Rather, a distinct set of facts anchors it.

Figure 6E shows the fact about collective acceptance at the lower left, anchoring the frame principle. The frame principle applies to all the possibilities in the frame. Whenever the grounding conditions are satisfied by some object, that object has the social property *being a dollar*.

---

[6] It may be more intuitive to understand anchoring as a function from sets of facts to frame principles, or else as a function from worlds (usually at times) to frames as a whole.

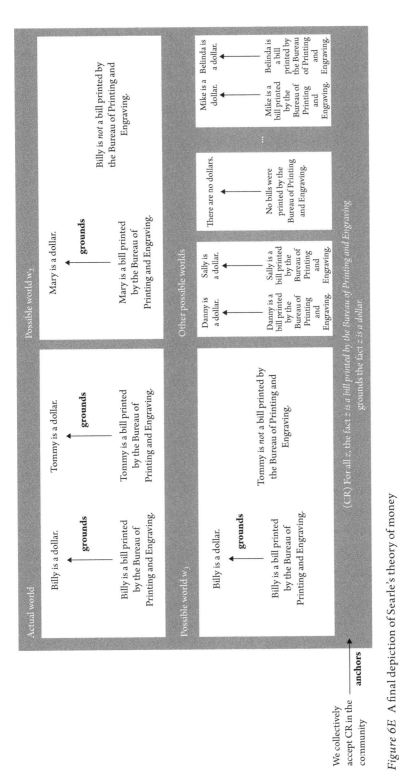

Figure 6E A final depiction of Searle's theory of money

This diagram is a useful one. But there is a lot going on, so it is easy to lose the point. To depict the overall framework and avoid confusion, then, I will usually leave out the other worlds (e.g., w_2, w_3, and the other worlds in figure 6E). But even when I simplify figure 6E it should be understood that a frame contains not just one world, but a universe of possible worlds, all anchored in the same way. That is, all conforming to the same frame principles.

A more general depiction of the framework is shown in figure 6F. In that diagram, it can be seen that there are two different places—highlighted in bold—where facts about individual people (and other facts as well) can play a role in "making the social world."

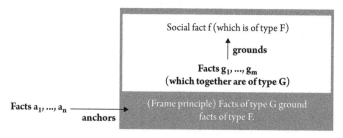

Figure 6F Generic anchoring and grounding diagram

In this figure, the **anchoring facts** $a_1, \ldots, a_n$ are the facts that set up or put in place the grounding conditions for social facts of type F. They are the facts, for instance, that set up the grounding conditions for *x is a dollar*, *x is a university*, *x is kosher*, *x is a murderer*, *x is a war criminal*, and so on. Facts about people may be involved in anchoring. For instance, the anchors may be the collective intentions of people in the community.

The **grounding facts** $g_1, \ldots, g_m$ are the facts that ground a social fact *f*. A grounded social fact in our frame might be one about a particular object, such as **Billy is a dollar bill**, **Tufts is a university**, or **Assad is a war criminal**. Or it might be a general fact, such as **There are dollars, I work at some university**, or **The International Criminal Court rarely punishes war criminals**. Any of these facts is grounded by a set of facts in the actual world. The same grounding conditions also apply to social facts in other possible worlds. For instance, there is a possible world in which Mother Teresa was a war criminal. In the actual world, she obviously did not satisfy the conditions for being a war criminal. But in some remote possible world, she committed such-and-such heinous acts, and hence satisfies those conditions we have anchored for *x is a war criminal*. Facts about people may be involved in grounding as well (or instead). For instance, Mother Teresa's acts may be the grounds for the social fact **Mother Teresa is a saint** or **Mother Teresa is a sinner**.

This fills out some general structure of the theories, of which Searle's and Hume's are examples. But there is still a loose thread. All the discussion so far has been focused on what I called the "Type 2 examples," ones in which the role of people in making social facts is that they put in place the grounding conditions. What about the relation between the mob and the mobsters? Here, the people are not putting in place the conditions for being a mob. The people *constitute* the mob.

Type 1 Examples

The bold type in figure 6F highlights two different roles individual people can play in "making" the social world. Social facts can be grounded by facts about people, and frame principles can be anchored by facts about people. With this clarification, the Type 1 examples—such as the mob, the flow of commuters, and the Jewish people—are simple to interpret. For these examples, facts about individual people play an exhaustive role, or at least a very significant role, in *grounding* certain social facts about them.

Recall the discussion of grounding from the last chapter. Facts (5.3) and (5.4) are two facts that stand in the partial grounding relation:

(5.3) **Bob, Jane, Tim, Joe, Linda, ... and Max ran down Howe Street**.

(5.4) **The mob ran down Howe Street**.

As I pointed out in the last chapter, (5.3) partially grounds (5.4), but does not quite fully ground it. For that, we need other facts, such as that Bob, Jane, and the others are clustered reasonably tightly together.

To work out what facts fully ground (5.4), one approach is to think more generally about the grounding conditions of a variety of kinds of facts. What, for instance, are the grounding conditions for a fact of the form *x constitutes a mob*? What are the grounding conditions for a fact of the form *x ran down Howe Street*? This is not a trivial project, even for so simple a fact as (5.4).

But even before we embark on such a project—and that will have to wait until Part Two—we can already see that these facts conform to the framework above. To ask about the grounding conditions for facts of the form *x constitutes a mob* is precisely to ask about the frame principles for facts of that form. In other words, a fact like **The mob ran down Howe Street** fits into the grounding and anchoring framework, just where any other social fact, such as **Billy is a dollar**, does. The only difference is that a fact like **The mob ran down Howe Street** is grounded by different sorts of facts than those that ground **Billy is a dollar**.

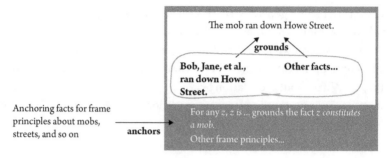

Figure 6G A Type 1 fact in the anchoring and grounding framework

In figure 6G, a circle is drawn around the facts that ground (5.4). Interestingly, this immediately raises a question that is scarcely noticed. What sorts of facts anchor the frame principles for facts about mobs? What is the "glue" holding together the social kind *mob*? Even if the Standard Model is correct for the case of dollars, or for the case of promises, it is not obvious that it is correct for the kind *mob*.

The Grounding Inquiry and the Anchoring Inquiry

For each of the two relations—grounding and anchoring—there is a separate project in social ontology. The grounding inquiry tends to precede the anchoring inquiry. It is the project of working out the frame principles in a given frame, usually our current frame, as it is actually anchored. The anchoring inquiry, on the other hand, examines how our frame principles are anchored, how we can anchor new frame principles, and how we can change frames.

In each inquiry, there are both specific and general questions to ask. For grounding: How is one particular social fact actually grounded? What are the different ways that particular fact might be grounded? And how are social facts of some *kind* actually grounded? What are the grounding conditions, in general, for facts of that kind? What are the grounding conditions for facts *in general*? Is a comprehensive set of the individualistic facts that obtain enough to ground all the social facts that obtain?

Similarly for anchoring. How is one particular frame principle in our frame anchored? How else might the same frame principle be anchored? What about different *kinds* of frame principles, or frame principles in general? Are there different *schemas* for anchoring frame principles? Are all the anchoring facts individualistic?

The anchoring and grounding model is meant to raise and clarify these questions. The overall structure of the model is not committed to any theory of anchoring or grounding. The structure of the model alone, however, already clarifies the difference between two different kinds of individualism about social ontology. There is individualism about grounding, which we already know by the name 'ontological individualism'. And there is individualism about anchoring. In chapters 8 and 9, I turn to those individualisms, both to clarify them and to elaborate on the tension between them. I also return to the distinction between anchoring and grounding, and show that an advocate of either individualism should be an even stronger proponent of the distinction between the two than the skeptic should be. First, though, I will say more about frame principles, both to avoid any misunderstanding of this key element of the model, and to exhibit a case in which even this sketch of a framework pays off.

7

Case Study: Laws as Frame Principles

The grounding–anchoring model has many parts: social facts, the grounding relation, grounding facts, frames, frame principles, the anchoring relation, and anchors. It is time for a concrete application, to see how these things work in practice. Most importantly, I want to clarify the notion of a frame principle, and what role it plays in the model.

An illuminating example, to do this, is the law. Laws can be understood as frame principles. They give the grounding conditions for certain social facts, and they have anchors.

Understanding laws as frame principles helps clarify both the grounding–anchoring framework and also the nature of law. It sheds light, for example, on the difference between the law and the law code—that is, the written documentation that tries to record what the law is. And it also helps clarify different theories of the "sources" of the law. All this is a nice immediate payoff of the model.

In this chapter I mostly discuss H. L. A. Hart's theory of law, laid out in his 1961 classic *The Concept of Law*. I show how it fits into the grounding–anchoring model, and how it illustrates the way frames can be "nested" within other frames. There are two different kinds of laws in Hart's theory—what he calls "primary rules" and "secondary rules." Each of these can be understood as frame principles, with different kinds of anchors.

Hart's Theory of Law

In *The Concept of Law*, Hart laid the groundwork for a new understanding of the nature of law. Hart's book recast and defended "legal positivism," and its framework became the structure for subsequent approaches to positivism and its opponents.

Theorists of the nature of law divide into two rough camps: the legal positivists and the natural law theorists. Legal positivists hold that facts about the law are facts about sociology, not morality. What counts as a law does so in virtue of facts about our social beliefs and practices. This means that there can be grossly unjust laws. What the law is is one thing, and what the law should be is another.

Natural law theorists, on the other hand, hold that an unjust law is no law. We have a duty to follow the law, but cannot have a duty to be immoral. These two rough approaches have ancient roots, and over the generations have been refined into a variety of nuanced positions. Hart's work influenced the course of these positions, not just among legal positivists, but among natural law theorists as well.

Hart divides the law into primary rules and secondary rules. Primary rules are those that direct and appraise behavior. Consider, for instance, the law on first-degree murder. The Massachusetts General Law (MGL) lists the following conditions:

> Murder committed with deliberately premeditated malice aforethought, or with extreme atrocity or cruelty, or in the commission or attempted commission of a crime punishable with death or imprisonment for life, is murder in the first degree. (MGL c. 265 §1)

The following section lists the punishments:

> Whoever is guilty of murder committed with deliberately premeditated malice aforethought or with extreme atrocity or cruelty, and who had attained the age of eighteen years at the time of the murder, may suffer the punishment of death pursuant to the procedures set forth in sections sixty-eight to seventy-one, inclusive, of chapter two hundred and seventy-nine. Any other person who is guilty of murder in the first degree shall be punished by imprisonment in the state prison for life. (MGL c. 265 §2)

These are all statements of primary rules. They involve the conditions for a certain kind of legal attribute, and given that, further conditions for sanctions.

Secondary rules identify, change, and enforce the primary ones. These govern how primary rules themselves are put into place and enforced. The most important secondary rule in a legal system is the system's "rule of recognition." This determines which other rules are recognized as legally valid, that is, what the primary rules are, in that legal system. Part of the rule of recognition in a modern legal system, for instance, pertains to how the legislature enacts statutes. In Massachusetts, these are set out in Articles I and II of the State Constitution. Another part of the rule of recognition is the role judicial interpretation plays in shaping the law. Other secondary rules in a system include the rules under which other rules may be changed, and the rules for who has the power to apply them.

The crux of Hart's argument for positivism is a theory of what puts secondary rules in place in a society. Hart gives sociological—and nonmoral—conditions for a secondary rule to be in place. Two conditions must be satisfied. First, legal practitioners, such as judges and lawyers, must have a convergent set of practices or behaviors. That is, they have to actually conform to the rule, when they act. Second, they all must take a certain attitude toward their practice.

More specifically, according to Hart, a rule R is present among a group P whenever there is a regularity in behavior such that: (1) most people in P regularly conform to R, and lapses from conforming to R are criticized; and (2) R is "accepted" in P, that is, R is treated as a standard for the behavior of people in P, and the criticism for lapses is regarded as justified.[1] These facts put secondary rules in place in a legal system, and the secondary rules contain the conditions for a primary rule to be in place.[2]

This is only a cursory description of primary and secondary rules, and I will fill out some details as I go. But even from this much, it may start to be clear how the grounding–anchoring model will apply. When we speak of "putting rules in place," we are speaking about anchoring. And the rules themselves give the grounding conditions for particular legal facts.

A Primary Rule as a Frame Principle

I will start with in the same way I began discussing grounding and anchoring: with a particular fact. Consider, for instance, the fact **Whitey Bulger is a first-degree murderer.** (In case you're not from Boston, Whitey Bulger is the famed local mobster who was recently captured and convicted, after a sixteen-year manhunt.) This legal fact is the case because of things Whitey Bulger actually did. It is grounded by historical facts about Whitey: facts about his having killed people and facts about his mental state in the course of doing so.

---

[1] Hart 1961, 55–60. See also Coleman and Leiter 1996; Green 1996; Greenberg 2004; Marmor 2010, 2012; Raz 1975; Shapiro 2011.

[2] Strictly speaking, Hart's secondary rules are a mix. The rule of recognition is put in place by these two kinds of facts, and other secondary rules (understood as rules about rules) are put in place the way primary rules are. That is, by satisfying the conditions set out in the rule of recognition. In this discussion, I will focus on the rule of recognition in speaking about secondary rules. I am grateful to Simon May for clarifying these points, and for detailed discussion of Hart and the law.

Massachusetts law contains specific conditions for having the property *being a first-degree murderer*: if someone kills a person with deliberately premeditated malice aforethought, then the killer is a first-degree murderer.[3] All it takes for someone to be a first-degree murderer is that the person satisfies those conditions. Since Whitey murdered several people, there are many facts that make it the case that Whitey Bulger is a first-degree murderer. Among them, for instance, is the fact **Whitey Bulger killed Bucky Barrett with deliberately premeditated malice aforethought**.

The connection between these facts is, of course, not just happenstance. The fact that Whitey killed Bucky *grounds* the legal fact about Whitey: it is the metaphysical reason for Whitey being a murderer. This means that a careful statement of the law will look similar to the frame principles I discussed in the last chapter. That is, something closer to: *For all x, if x kills a person with deliberately premeditated malice aforethought, then that grounds the fact that x is a first-degree murderer.*

This is just one example of a primary rule. Laws come in many forms. It is typical, for instance, for some primary rules to set out the conditions for a person to have some legal attribute, like *being a murderer*. And for other primary rules to set out the sanctions that accompany that attribute. We saw this in the two quoted sections of the Massachusetts General Law. Moreover, not all primary rules have the form *For all x, if x has property P, then that grounds the fact that x has property Q*. In many cases, laws just give the conditions for grounding particular facts, not kinds of facts. They are also sometimes categorical facts, such as *A has property Q*, as opposed to conditionals. (A nice thing about using the law as a case study is that it provides a wealth of examples for broadening the forms of frame principles.)

What Exactly Is the Legal Code?

The passage from the Massachusetts code quoted above lists the conditions for murder in the first degree. This passage does a reasonably good job capturing the conditions for being a first-degree murderer.[4]

This passage from the Massachusetts code, however, is not a perfect statement of the law—that is, the frame principle. The law is distinct from the legal

---

[3] I am leaving out a few conditions, as can be seen in the passage from MGL. Also, certain exceptions are made for legal killing.

[4] There is, of course, a difference between being a murderer and being found guilty of murder. The latter depends on people's judgments, and the former does not.

code, although the legal code plays important roles in the law. To see what the legal code does, we need to distinguish anchors from frame principles.

One issue with the passage I quoted (i.e., MGL c. 265 §1) is the one I mentioned a moment ago, about grounding. On the surface, this passage looks as though it is an identity statement. It says of such-and-such behavior that it *is* first-degree murder. But statements like these are better understood as giving the grounding conditions for having a social property. (Searle rightly stresses that *being a first-degree murderer* is a status. Statuses are not identical to their conditions. Being a dollar is not the same thing as being printed at a certain bureau.) Thus this passage does not quite capture, or at least does not explicitly capture, what we want: the grounding connection between facts.

Interestingly, even apart from the issue of grounding, the Massachusetts law is *still* not exactly what is written in the code. First, although the legal code is an attempt to record the law, it is not a definitive attempt. The law can diverge from what is written in the legal code. Second, the legal code itself is a part of what anchors the law. I will explain these two roles, and then it will be easy to identify them in a grounding–anchoring diagram.

1. The Divergence between the Law and the Legal Code

The obvious source of divergence between the code and the law is that Massachusetts is a common-law jurisdiction. This means that judges are bound by precedent, and thus that historical decisions by the judiciary anchor, in large part, the law itself. If there is a judicial tradition that diverges from the enacted statutes, that tradition carries weight in determining what the law is.

In addition, the legal code is also distinct from the statutes in force. The Massachusetts code, for instance, is the best attempt of the House and Senate Counsel to represent the law. But not everything that is enacted ends up in the code. The courts routinely rely on unencoded legislation. Moreover, not everything in the code is formally enacted. And even the laws that are not enacted have force in determining what the law is.

Finally, you might notice that the passage above does not actually spell out the conditions for *being a murderer*. It gives the conditions for *being a first-degree murderer* in terms of *being a murderer*. But the legal code lacks further specification of the legal conditions for *being a murderer*. Those conditions are something else that is anchored in part by practices in the judiciary. Interestingly, to track down the anchors for these conditions, we must look to things like the instructions that judges give to juries in homicide cases.[5]

---

[5] See, for instance, http://www.mass.gov/courts/docs/sjc/docs/model-jury-instructions-homicide.pdf.

2. The Legal Code as Partially Anchoring the Law

None of this means that the legal code itself is merely an inert, imperfect record. Rather, the legal code carries weight in anchoring what the law is. What is actually written in the legal code, in other words, is a part of the overall package that anchors the law.

It makes sense that the legal code should play these two roles. It is a practical document, an instrument for applying the law. In applying the law, we are interested in knowing the legal facts that obtain. Is Whitey Bulger a first-degree murderer, or not? Under what conditions does a particular legal fact obtain? Is it legal to avoid taxes in such-and-such a way, or to copy such-and-such a document for personal use? These questions concern facts in our current frame. They are questions about the law as it has been anchored. The anchoring is complex, and in order to work out what the law is, we sometimes have to go back and look at the whole network of interacting anchors. But most of the time that complicated network is largely superfluous. Most of the time, we can just look at what we have recorded—the legal code itself—as our best stab at the grounding conditions. That is close enough.

That explains why the legal code is an imperfect record of the law. But why should the written code itself be among the anchors for the law? Because the legal code is part of the network of legal practice. Much of what is important in anchoring the actual law is what elements of the legal system do. And just as lawyers and judges are legal participants, the code is effectively a legal participant as well. Thus the code itself carries weight not just as a fallible record, but as a part of what puts the law in place.

The Grounding–Anchoring Diagram

Figure 7A is a rough picture of how **Whitey Bulger is a first-degree murderer** is anchored and grounded. Grounding that fact is the fact **Whitey killed Bucky Barrett with deliberately premeditated malice aforethought**. The conditions for being a murderer are facts about the law, that is, the frame principle for facts of the form ***x* is a first-degree murderer**. MGL c. 265 §1 is the House and Senate Counsel's attempt to record this law or frame principle. The law, in turn, is anchored by a variety of facts, including the history of trials and judicial decisions, actions of the legislature and the governor, and what the counsel in fact recorded in the MGL.

It makes sense that we tend to think of the legal code as just being the law, even though it is not. After all, for simple things like first degree murder, the legal code does a reasonable job expressing the law. It is easy to overlook that the legal code is an imperfect statement of the law, and even easier to

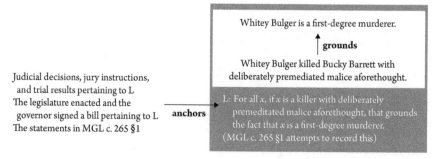

Figure 7A Anchoring a law

overestimate the anchoring role the legal code plays. But it is not an either-or choice: that the law is anchored exclusively by the text, or not at all. That false choice only reflects an impoverished theory of anchoring.

Secondary Rules and Their Anchors

The bottom of figure 7A represents a *particular* anchoring. One particular set of anchors—a particular set of legislative actions, judicial decisions, jury instructions, the legal code, and so on—anchors a particular primary rule.

This anchoring is just one instance of a more general rule: a secondary rule. The rule of recognition gives the conditions for a primary rule to be legally valid. That is, it gives the anchoring conditions for a primary rule.[6] The rule of recognition, then, is the general statement of which the bottom of figure 7A is an instance.

The Rule of Recognition

A rule of recognition needs to adjudicate among competing factors, in anchoring a valid law. Many factors are involved in this anchoring, including court hierarchies, various branches of the legislature and the executive, legal codes, and so on. Hart takes there to be one unified rule of recognition for a legal system, but we should expect this to have an extraordinary complicated antecedent. The bottom of figure 7A only gestures at the anchors that put law L in place.

Still, even without worrying about all the complexities, we can say a bit about the form of a rule of recognition. Start, for instance, with a statute enacted by

---

[6] Greenawalt 1986; Shapiro 2009; 2011, 84–6.

a legislature. As I pointed out, that statute is not the law. So in the statement of the rule of recognition, we need to distinguish these. A rule of recognition may have something like the following form:

(7.1) For all y, if a statute y' is enacted by the legislature and signed by the governor, and y is the product of y' together with the relevant influences of judicial decision, jury instructions, trial results, and statements in the legal code in such-and-such a way, then together those facts anchor y.

In this statement, I distinguish the law y from a statute y'. The idea is that y' is but one component of the anchors of a law y, leaving room for the other anchors to have their effects in molding y. (Of course, this statement of the rule is still just a gesture. It also assumes that laws originate in statutes, which is not the way much of common law works.)

Anchoring the Rule of Recognition

As Hart recognized, the key question for the nature of law is not so much what the rule of recognition is, but what puts it in place. How, in other words, is this rule anchored? Here is where Hart's "theory of practices" comes in.

The theory of practices is Hart's theory of the conditions under which the rule of recognition is in place in society. The two conditions are: (1) the conformance of behaviors and practices to the rule; and (2) the acceptance of the rule among the practitioners. This theory is closely related to Hume's earlier theory of convention (and to later versions, such as David Lewis's), and also to Searle's subsequent theory of collective acceptance. Hart's theory is another variant on the "Standard Model of Social Ontology." Like Searle, Hart requires that members of the community have attitudes toward the secondary rule itself. Hart partly takes secondary rules to be anchored by attitudes involving R as a standard of behavior. And like Hume, Hart also adds explicit conditions on the actual behaviors: the behaviors must be regularly conformed to, and lapses actively criticized.

Just like these other views, we can depict Hart's theory of practices as a component of the grounding–anchoring model as a whole. This is shown in figure 7B. At the left side of the figure are the anchors, according to Hart's theory. That is, the facts about convergent practices and attitudes regarding R. These facts anchor the frame principle R, the rule of recognition.

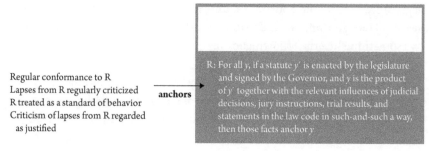

Figure 7B Hart's anchoring of a secondary rule

Putting the Pictures Together

There is, of course, a relation between figures 7A and 7B. Figure 7B is different from all the other figures I have drawn, in one important way: what is anchored is not a frame principle governing the grounding of particular facts, but rather a frame principle governing the anchoring of frame principles. The anchoring depicted in figure 7A is governed by the frame principle anchored in 7B. That is, figure 7A is nested within the frame depicted in 7B.

This is what Hart is getting at in distinguishing primary and secondary rules. The secondary rules are the ones that set up the law-making principles for the primary rules. Primary rules are enacted within a frame governed by secondary rules. When, for instance, a law like *killers with deliberately premeditated malice aforethought are first-degree murderers* is enacted, that involves the lawmakers satisfying the grounding conditions expressed in the secondary rule. Those secondary rules are, in turn, anchored (in Hart's view) by the facts discussed in his theory of practices. Hart's theory, modeled as a nested grounding–anchoring model, is depicted in figure 7C.

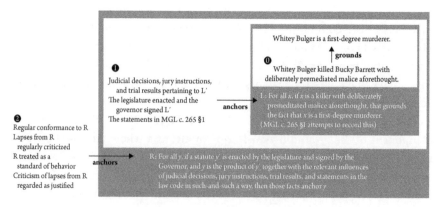

Figure 7C Nested frames depicting the anchoring of secondary and primary rules

Again, on the far left of the diagram (marked as ❷) are the "practices"—the convergent behaviors and attitudes toward those behaviors. Together, those facts anchor the rule of recognition. That rule gives the criteria for legal validity, that is, the facts that must obtain in order for a law to be anchored in the frame.

As I did in chapter 6, figures 6F–6G, I have left out all of the other possibilities in the frame. (Imagine how complicated the diagram would be if other possibilities were also depicted.) Still, it should not be forgotten that the secondary rule expresses conditions for any possible set of lawmaking facts in the frame. In a possibility where the relevant parties go through the relevant steps to enact and mold A, that anchors the fact that A is a law; and in a possibility where the relevant parties go through the relevant steps to enact and mold B, those steps anchor the fact that B is a law.

Now looking within the large frame, we see one actual set of facts (❶) anchoring a primary rule. These facts include the actions of the Massachusetts legislature and the governor, and also judicial practices, as well as facts about what is actually recorded in MGL. The rule of recognition governs the fact that these factors anchor the law on first degree murder. Then within the small frame is the actual grounding fact about what Whitey Bulger did (❶), and the legal fact that his actions ground.

In some ways it is a complicated picture, but overall a fairly straightforward framework. One thing it clears up is the essential distinction between a frame principle and a set of anchors. This seems obvious, but is frequently elided. David Hume and David Lewis, for instance, often use the term 'convention' to refer both to the anchoring facts and to the frame principle. Likewise, Hart is vague about the application of the term 'rule', sometimes using it to refer to the regularities in behavior and attitudes. These are not the rule, but the rule's anchors, that is, the reasons that the rule is in place.

As I mentioned, the model also helps clarify some important distinctions in the subsequent literature on legal positivism. We can quickly use it to pick out various positions.

The Hart-Dworkin Debate

Legal positivism holds that laws are legally binding in virtue of their social pedigree. It holds that the moral content of law is not relevant to its being legally binding. Not long after Hart's book was published, Ronald Dworkin began a series of critiques in which he argued that moral considerations are indeed relevant to a law's being legally binding.[7] Judges routinely invoke moral considerations in their decisions, Dworkin points out. For instance, they overturn

---

[7] Dworkin 1978, 1986.

contracts on the basis of fairness. Consequently, says Dworkin, moral considerations figure into what the law is. Some philosophers rejected this, insisting that where judges brought in moral considerations, they were straying outside the boundaries of the law. This view has come to be known as "exclusive" legal positivism. Other philosophers, however—Hart included—replied that positivism can accommodate moral criteria for legal validity. It is not the criteria of legal validity that need to be strictly social and free of moral considerations. Rather, it is the *source* of those criteria. This view has come to be known as "inclusive" legal positivism.

A long description of the distinction between exclusive and inclusive legal positivism is unnecessary, since it can easily be shown using figure 7C. The differences between exclusive legal positivism, inclusive legal positivism, and various versions of antipositivism, are differences among what theorists take to be the anchors for primary and secondary rules.[8]

In figure 7C, ❶ marks the anchors of the primary rules, and ❷ marks the anchors of the secondary rules. Views on the sources of legality can be distinguished by their different commitments to what sorts of facts can be included in these categories. Different views are listed in table 7a, along with what they take to be the anchors of primary and secondary rules.[9]

Table 7a **Theories of the "sources" of law**

	❶ Anchors of primary rules	❷ Anchors of secondary rules
Exclusive legal positivism	Strictly social	Strictly social
Hart's version of inclusive legal positivism	Mixed (no restrictions)	Convergent beliefs and practices about the rules
Inclusive legal positivism	Mixed (no restrictions)	Strictly social
Dworkin's "legal interpretivism"	Both social and moral facts	Both social and moral facts
Certain religious natural law theories	Views differ on these	Divine sources, basic human purposes

[8] Leiter 2003 and Shapiro 2007 are excellent presentations of these debates.
[9] It does some violence to Dworkin's view to place it in the table, since he rejects the distinction between primary and secondary rules.

Exclusive legal positivism insists that morality has no place in anchoring laws, either in the criteria for legal validity (i.e., in the anchors for the primary rules) or in the sources of those criteria (i.e., in the anchors for the secondary rules). When a judge invokes moral considerations in determining whether a law is legally valid, the judge is using moral facts in category ❶. This, according to the exclusive legal positivist, is an extralegal move.

Inclusive legal positivism allows moral considerations to enter into the determination of legal validity. Although the *secondary* rules are anchored by strictly sociological facts, that does not exclude moral facts from being anchors of *primary* rules. It may be our practice, for instance, to allow judges to weigh moral considerations in determining whether a law applies to a particular case. That allowance is part of the content of our rule of recognition because it is our common practice. The anchors of the secondary rules remain strictly social, despite the anchoring conditions in the secondary rules themselves having moral content.

Whichever position turns out to be most attractive, in large part the debate can be seen as one about anchoring. Theories of the criteria of legal validity are theories of the sorts of facts that anchor primary rules. And theories of the sources of legality are theories of the sorts of facts that anchor secondary rules.[10]

Model Building and the Grounding Inquiry

In practice, most lawyers and judges are not concerned with the anchoring inquiry, that is, working out the criteria for legal validity or the sources of those criteria. Nor are most lawyers and judges concerned with the grounding inquiry, that is, working out what the laws (the frame principles) are. Mostly they are interested in the determination of fact. They take the law to be understood. Their work is to establish whether the particular grounding facts obtain. In the Whitey Bulger trial, for instance, there was no discussion about the conditions for being a first-degree murderer. The question was whether Whitey satisfied those conditions. Did he kill? Was it with malice aforethought? This is neither the anchoring inquiry nor the grounding inquiry. It is what we might call the "actual fact" inquiry.

A related part of the practice of law is concerned not with the determination of actual facts, but instead with exploring possible facts. Much of the practice of tax law, for instance, is concerned with exploring different possible

---

[10] A recent proposal is Shapiro 2011.

structures, to see whether they satisfy the conditions for being taxable. When Verizon bought out the stake held by Vodaphone, was that a taxable transaction? What if they had structured the transaction in a different way? This is an investigation of possible facts, and the application of frame principles to evaluate whether various other facts would ground the possible legal fact **The buyout is taxable**. This project we might call the "model building" project.

However, to do either of these projects well—to determine the actual facts or to build models of possible facts—we need to have a good account of what the laws are. That is, for *somebody* to have done the grounding inquiry well.

For legal "model builders," the anchoring inquiry is not usually relevant. They are simply interested in having people get the law in our frame right—that is, the grounding inquiry—and in using the relevant frame principles as the basis for thinking through possibilities. Only in certain cases would a model builder be interested in the anchoring inquiry. For instance, if someone were interested in modeling the effect of lobbying efforts on various legal participants, then she would need to consider how those participants figure into anchoring the law.

The same point applies to frame principles in general, and to the social sciences in general. The bulk of projects in the social sciences are "actual fact" inquiries and "model building" projects. This is why social scientists are often familiar with individualism of the sort discussed by Popper and Watkins, and not with the "Standard Model" and its advocates, such as Searle. When the aim is to determine what the actual social facts are, or how they change with changes in the grounding facts, what matters is getting the grounding conditions right. For most modeling it is more important to know what the frame principles are than to know what sorts of facts put the frame principles in place.

In the next chapter, I turn to the distinction between ontological individualism—that is, individualism about the grounds of social facts—and individualism about the anchors of social facts. Subsequently, I return to the reasons for making a sharp distinction between grounds and anchors, rather than just seeing anchors as a kind of ground.

8

Two Kinds of Individualism

In chapter 3, I raised an alarm about ontological individualism. Ontological individualism has long been regarded as obviously true. But it may be no more defensible than Virchow's extreme version of cell theory.

All along, however, there have been two traditions about the nature of the social world. The ontological individualism tradition of Watkins, Lukes, Kincaid, and Pettit is one. The "Standard Model" tradition of Hume, Hart, and Searle is another. In the last few chapters, I have assembled a new model for making sense of both, unifying them in a single framework.

What does this mean for individualism about the social world? How should this change our understanding of the original thesis of ontological individualism? And how should we construe individualism in the "Standard Model" tradition? With the model and tools in hand, we can resolve these. There are two distinct kinds of individualism, two different kinds of claims about how social facts are made by facts about individuals. These correspond to two different locations in the model, as shown in figure 8A.

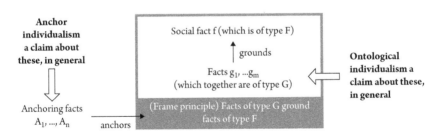

Figure 8A Two kinds of individualism

What I will call "anchor individualism" is a claim about how frame principles can be anchored. Ontological individualism, in contrast, is best understood as a claim about how social facts can be grounded. I will begin with a comment on the sets of social and individualistic facts, and then turn to a brief sketch of anchor individualism. Most of this chapter, however, is devoted to revisiting ontological individualism. Moving beyond the "supervenience" interpretation, I introduce two complementary notions: *determination* and *dependence*. Both of these can be analyzed in terms of grounding, and with them, we can develop a more up-to-date understanding of ontological individualism. Supervenience turns out to be a useful (albeit imperfect) tool for evaluating whether ontological individualism is true. But it is not the best way to analyze ontological individualism itself.

I will defer one important issue to the next chapter: that anchors are not included among the grounds of a social fact, and correspondingly that ontological individualism is not about anchoring.

Individualistic Facts

To put forward any version of individualism, a key task is to identify which facts are the social facts, and which facts are the individualistic ones. This task is unavoidable for a proponent of individualism. The individualist needs to have a clear sense of these. Otherwise, it is pointless for her to assert that the social facts are exhaustively "built out of" the individualistic ones.

I am not confident this can be done: debate about these categories has been going on for decades. For this reason, I want to avoid evaluating the dozens of conceptions of "individualistic" and "social."[1] Instead, I will *grant* to the individualist that we can take each proposition and sort it into one of three mutually exclusive categories: the social, the individualistic, and the ones that are neither social nor individualistic. As I discussed in chapter 5, the propositions corresponding to the social facts need to be understood more broadly than they often are. If we can figure out which properties and objects are the social ones, the social propositions are plausibly the ones that have any social object or social property as a constituent. For instance, propositions like *The mob is cold* and *John is rich* are each social propositions. The individualistic propositions are those that have individual people or individualistic properties as constituents, but not social ones. *John is cold* is an individualistic proposition. Some propositions are neither social nor individualistic, such as *The sun is 93 million miles from the earth*.

The social facts, then, are those corresponding to the social propositions, and the individualistic facts are those corresponding to the individualistic

---

[1] I do discuss this in Epstein 2009 and Epstein 2014a.

propositions. I will use the letter S to denote the set of possible social facts, the letter N for the set of possible individualistic facts, and the letter Z for the set of possible facts that are neither social nor individualistic.

What Is Anchor Individualism?

Anchor individualism is a thesis about how frame principles are anchored. It is a thesis about anchoring in general: all frame principles, across all frames, are exhaustively anchored by facts about individual people. Searle's collective acceptance theory is an example. Whenever a constitutive rule is anchored in a community, it is anchored by the "we-attitudes" of individuals in the community. Searle makes few commitments to what sorts of facts ground social facts. But his "collective acceptance" theory restricts anchors to a very limited set of mental states of individual people.[2]

If Searle is right, that means that every actual frame principle is anchored in the same way. And that every *possible* frame principle is also anchored in the same way: always by collective acceptance of the principle itself.[3] To illustrate, table 8a lists the frame principles and anchors for two different frames. The "Frame 1" column lists frame principles for ***x is a dollar*** and ***x is a first-degree murderer***. And the "Frame 2" column gives a different set of frame principles for these kinds of social facts. Each also lists the respective set of anchors, according to Searle's theory.

Table 8a **Multiple frames**

	Frame 1	Frame 2
Frame principles	• (D1) If x is a bill issued by the Bureau of Engraving and Printing, that grounds the fact that x is a dollar. • (M1) If x kills with deliberately premeditated malice aforethought, that grounds the fact that x is a first-degree murderer.	• (D2) If x is a bill issued by the Treasury Department, that grounds the fact that x is a dollar. • (M2) If x kills deliberately but without malice, that grounds the fact that x is a first-degree murderer.
Anchors	• We collectively accept D1 • We collectively accept M1	• We collectively accept D2 • We collectively accept M2

[2] In Searle's view, a "we-attitude" is simply a different kind of internal mental state of an individual than an "I-attitude" is (Searle 2010, 42–50).

[3] Searle has slightly modified this in Searle 2010, with his discussion of collective recognition.

Other Versions of Anchor Individualism

Searle's theory is not the only version of anchor individualism. All the versions of the Standard Model of Social Ontology are also varieties. There may also be versions of anchor individualism that are not representatives of the Standard Model at all.

One way that Searle's theory is particularly restrictive is in its limited set of anchors. Only attitudes toward frame principles—not more general attitudes, not regularities, not practices, nor anything else—anchor frame principles themselves. The other theories I have mentioned allow for more inclusive anchoring facts. On Hart's theory of practices, the anchors of a rule R are not only attitudes toward R, but also behaviors conforming to R. In other words, the anchors involve actions, not just attitudes. Hume's theory is slightly more liberal still. It involves attitudes and behaviors as well, only the attitudes do not need to be attitudes toward the convention. Instead, the attitudes are about the sorts of things that other members of the community know and want.

Of the three, only Searle's version holds that frame principles are anchored by psychological facts about individual people. Despite this, in all the accounts, the facts anchoring frame principles are facts about individual people. The theorist who denies anchor individualism will take anchors to include facts that are *not* facts about individual people.

Anchors for Each Frame Principle, or Just for the Whole Set of Them?

Searle's theory is also particularly restrictive in that it proposes that any given constitutive rule have a *specific* anchor. Each rule has one corresponding set of anchoring facts, and in Searle's theory, the corresponding set is easy to identify: *Constitutive rule x holds in a community c if and only if every member of c has the we-attitude: We accept x.*[4]

This sort of correspondence, however, is more than the anchor individualist needs. We could construct a less reductive thesis of anchor individualism, using the same kind of generality that I described in connection with ontological individualism. As I discussed in chapter 2, ontological individualism says only that the social facts are exhaustively determined by individualistic facts *as a whole*. It is silent on whether or not a reductive account of the grounds of social facts is possible. To be ontological individualists, we do not need every

---

[4] Searle's theory is also narrow in its formulation of frame principles, as *X counts as Y in C*. But this is a matter of what he takes to be anchored, not the facts that do the anchoring, which is what anchor individualism pertains to.

social property to be connected to a specific set of individual level properties. Similarly, a theorist who argues that anchors in general must be individualistic does not need to specify the *particular* anchors for any given frame principle. Instead, he may just make a claim about the relation of a whole set of anchors to a whole set of frame principles.

A theory of convention, for instance, may take there to be a diverse set of facts that together anchor a variety of conventions. It may not regard conventions as being anchored one by one. It may be, for instance, that we have a huge number of conventions about driving cars, and that all of these conventions are jointly anchored by a huge set of interconnected behaviors and beliefs about driving. Just as ontological individualism is the generic thesis that social facts are exhaustively built out of facts about individuals, anchor individualism can be the generic thesis that the grounding conditions for social facts are exhaustively anchored by facts about individuals. That is, it is a claim about a relation holding between a whole set of "social-level principles" and a whole set of "individual-level facts."

In short, one could be an anchor individualist without having a specific theory of how individual frame principles are anchored. One merely needs to be committed to the view that frame principles in general are anchored by facts about individuals.

The Anchoring Inquiry

Theories about how we anchor the social world arguably stretch back to Plato and Aristotle, both of whom discussed the role of agreement in forming language and laws.[5] Despite this long tradition, there are still surprisingly few theories of how and why certain facts anchor certain kinds of rules or principles. These theories have been stunted for a number of reasons: We often fail to distinguish anchors from grounds, so the inquiries have become muddled; theorists have been unwilling to look deeply into the nuances of individual cases; the relation between convention and collective intentions has remained murky; and most importantly, theorists have insisted that there can be only one schema by which anchoring works.

Even a quick look at the last chapter shows that this is unlikely to be right. Hart's theory of practices, for instance, is only meant to apply to the rule of recognition. A theory of anchoring needs to confront the frame principles for many kinds of social facts. Frame principles for facts of the form ***x is a murderer*** or ***x is a felon*** may be anchored in the way that primary rules are. Those for facts of the form ***x is a US law*** may be anchored in the way secondary

---

[5] See, for instance, *Cratylus* 384d and *De interpretatione* 16b19.

rules are. Those for facts of the form *x is money* or *x is a corporation* may be anchored still differently.

Altogether, the anchoring inquiry is wide open terrain, both for anchor individualists and for their opponents. In my view, the best way to assess anchor individualism is to consider anchoring more generally. And that is a topic for a bigger book than this one.

As I pointed out, however, the "grounding inquiry" is more pertinent to modeling than is the "anchoring inquiry." This means that, at least for the practice of social science, the failure of ontological individualism has a more immediate impact than the failure of anchor individualism does.

How then, after all this, should we understand ontological individualism? It is almost always analyzed in terms of supervenience, but that is not the best way to go. In the next sections, I develop a new analysis. Ontological individualism is better understood as a thesis about grounding. Or more specifically, as a thesis about dependence, which I will define in terms of grounding.

I will then return to supervenience, to explain two points. First is the relation between a dependence claim and a supervenience claim. Second, and much more important, is the common mistake in applying supervenience to facts.

Making Sense of Determination and Dependence

What does it mean to say that social facts *depend on* facts about individuals, or that facts about individuals *determine* the social facts? Dependence and determination differ somewhat. Both can be understood in terms of the grounding relation. Both are claims about how various possible facts can be grounded. First I will consider what it means to say that a fact G determines a fact F, and what it means to say that a fact F depends on a fact G. Subsequently I will turn to *sets* of facts.

The words 'determines' and 'depends on' are often used loosely. When we say that G determines F, we sometimes mean a cause-and-effect relation: G causes F to be the case. Other times, we mean a grounding relation: G is a full metaphysical reason that F obtains. Here I will use it with one precise and stronger meaning.

(8.1) *G determines F*: it is necessary that if G is the case, G grounds F.

You will recognize this as the form of many of the frame principles I have discussed above, with an added explicit "it is necessary that." For instance, suppose that G is the fact **Whitey killed Bucky with deliberately premeditated malice aforethought**, and F is the fact **Whitey is a first-degree murderer**. Then, to say that G determines F is to say *It is necessary that if Whitey killed Bucky with deliberately premeditated malice aforethought, that fact grounds the fact that Whitey is a first-degree murderer.*

In short, the idea of *G determining F* is that anywhere, in any possible world, whenever you have G, that is enough for it to fully ground F. This leaves open the possibility that F can be grounded in other ways. If you have G, then G fully grounds F, but you could have F without its being grounded by G. (Recall that *G grounds F* is the same as *G fully grounds F*.)[6]

Sometimes people use 'depends on' interchangeably with 'determines'. This is not quite right. There is a subtle but significant difference between the two.

One thing that makes the notions of determination and dependence confusing is that they take their relata in a different order. We say, for instance, that facts about my brain determine facts about my thoughts, and that facts about my thoughts depend on facts about my brain. To avoid the confusion, I will keep using the letters F and G in the same way I did above: G is the grounding fact, and F is the fact that is grounded. Again, we can take G to be a fact like **Whitey killed Bucky with deliberately premeditated malice aforethought**, and F to be a fact like **Whitey is a first-degree murderer**.

The intuitive difference between determination and dependence is this. To say that G determines F is to say that G always makes F obtain. More specifically: G being the case guarantees that F is the case, and G is a complete metaphysical reason for F. To say that F *depends* on G is to say that F needs G, in order for F to obtain. More specifically: F guarantees that G is the case, and that G is at least part of the metaphysical explanation for F. Thus there are two differences between determination and dependence. One is that dependence has F as the antecedent of its conditional. The other difference is that dependence need not involve the full grounding relation, but only involves partial grounding:

(8.2)   *F depends on G*: it is necessary that if F is the case, G partially grounds F.

If F depends on G, then anywhere, in any possible world, if F obtains, then it is partially grounded by G. But it leaves open the possibility that G can obtain without grounding F.[7]

---

[6] I mentioned in chapter 5 that I side with the "contingency" rather than the "necessitarian" view of grounding. That detail matters if we are to make sense of my definition of determination. The problem is that on the competing view, my definition is redundant — determination is implied by grounding. It would not matter terribly if it were redundant, but I want to be sure to avoid confusion. Since I do not assume that grounding implies necessitation, the definition of determination is not redundant.

[7] Interestingly, in the literature on grounding in metaphysics, there is an active discussion of dependence, but determination is somewhat overlooked. Rosen 2010, for instance, errs in his

Although I am reluctant to burden us with too many relations, I should introduce one more: the full dependence relation. That is the same as dependence, but with full grounding instead of partial grounding. This relation will be useful in a moment, when we start thinking about how sets of facts can depend on other sets of facts. So the definition of full dependence is this:

(8.3) *F fully depends on G*: it is necessary that if F is the case, G grounds F.

Why isn't this the intuitive notion of dependence, and why instead do I say that dependence just involves partial grounding? Because plain dependence just involves some reason, one of the reasons, for the fact obtaining. We say that **Whitey is a first-degree murderer** depends on the fact **Whitey killed somebody.** That is true, even though having killed somebody is not itself enough to fully ground being a first-degree murderer. F depends on G just in case it needs G. G may not be the only thing F needs. F *fully* depends on G just in case it is the *only* thing F needs.

Thus there is a sort of imperfect symmetry between determination and dependence. The determination relation between **Whitey killed Bucky with deliberately premeditated malice aforethought** and **Whitey is a first-degree murderer** is that whenever you have the killing, that entails that the murderer fact also obtains. The dependence relation is that whenever you have the murderer fact, that entails that it is partially grounded by the killing. Notice that in this example, determination holds but dependence does not. **Whitey killed Bucky with deliberately premeditated malice aforethought** determines **Whitey is a first-degree murderer**. But **Whitey is a first-degree murderer** does not depend on **Whitey killed Bucky with deliberately premeditated malice aforethought**. After all, suppose **Whitey hadn't killed Bucky**. He still killed lots of other people with deliberately premeditated malice aforethought. So he would have been a first-degree murderer without that fact being even partially grounded by the killing of Bucky.

Here is another example to illustrate determination and dependence. Consider the following three facts:

(8.4) *I married Sarah on July 13, 2008, both of us are alive, and we have not divorced.*

(8.5) *I married someone at some point in the past.*

(8.6) *I am married.*

account of "real definition," because he uses a formula like (8.3) and overlooks the need for a determination clause.

(8.4) determines (8.6): it is necessary that if I married Sarah in the past, and we are both alive and undivorced, then those facts fully ground the fact that I am married. But (8.6) does not depend on (8.4). I could have been married to someone else, so that fact could obtain without being even partially grounded by (8.4).

On the other hand, (8.6) does depend on (8.5): it must be that if I am married, that fact is partially grounded by the fact that I married somebody in the past. But (8.5) does not determine (8.6): even if I married someone in the past, I might have divorced in the interim.

Defining Ontological Individualism

Ontological individualism is not just a claim about one social fact and one individualistic fact. It is a claim about the grounding of all the social facts. Up to this point, I have spoken of it as a claim that social facts are exhaustively determined by facts about individuals. But ontological individualism is best understood as a claim about dependence—about the dependence of all social facts on some set of individualistic facts. The ontological individualist says that if we take the social facts that obtain, then the grounds for those facts—the full metaphysical reason for their obtaining—is some set of individualistic facts.

It will come as no surprise that there are various ways to cash this out. One natural way is to look at all the social facts in all the possible worlds. For each social fact, if that social fact obtains in a world, then also obtaining in that world is a set of individualistic facts that grounds that social fact.[8] That is:

> (OI1) For any possible world w, and any social fact f at w, there is some subset X of N (where N is the set of possible individualistic facts), such that X grounds f at w.

This is a dependence claim. All social facts in all worlds depend on some set or other of individualistic facts in that world.

This is not the only way to fill in the details. Here is one reservation we could have about this definition of ontological individualism: it takes each social fact to have its own individualistic grounds. This seems hard to deny, if one is to be an ontological individualist. But we could come up with a weaker interpretation, which does not require this and still seems to qualify as ontological individualism. That is, if we take the set of social facts in any world, that whole set taken together must be fully grounded by some set of individualistic facts

---

[8] In the next chapter, I discuss the restriction to our frame.

in that world. To put it more precisely, we need to refer to one huge fact we can call the "total social fact" for a world. Take all the social facts that obtain in a world—that is, the subset of S all of whose members obtain in the world. Then let the "total social fact" in that world be the single fact that all the facts in that subset obtain. Then ontological individualism can be understood as the following:

> (OI2) For any possible world w, if F is the total social fact at w, then there is some subset X of N, such that X grounds F at w.

We can also make other changes to account for different ways we might take the individualistic facts to relate to the social facts.[9] All these details show that ontological individualism is a family of theses, not just one. We can precisely define a variety of versions, using the grounding relation. These definitions capture its intuitive force, and also capture various nuances that the ontological individualist may or may not want to commit to.

These definitions also show that if we want to confront ontological individualism, either to support it or attack it, we do it via the grounding inquiry. What are the grounding conditions for social facts? What are, in other words, their frame principles?

And they show how we can distinguish ontological individualism from its traditional analysis as a supervenience thesis. Supervenience has its uses, but it also has significant limitations.

Supervenience

Ontological individualism is a claim about what grounds what, rather than a claim about what supervenes on what. Supervenience should be regarded as a family of diagnostic tools. The supervenience of one set of properties on another set of properties provides strong—but not infallible—evidence that the facts involving the first set metaphysically depend (in one way or another) on the facts involving the other set. But it only provides that evidence if it is correctly set up and interpreted.

I will consider only the one best candidate for formulating ontological individualism: global supervenience.[10] Global supervenience, like most forms, is a

---

[9] For instance, we can change 'grounds' to 'determines', to ensure that individualistic facts ground the social facts in a uniform way across all possibilities.

[10] There are, in fact, several versions of global supervenience, but their differences are not important for our purposes; cf. Epstein 2009.

relation between sets of properties: a set A of properties and a set B of properties. Intuitively, to say that *A globally supervenes on B* is to say that if we fix all of the B-properties in the whole world, then that suffices to fix all the A-properties in the world as well. To say that the *chemical properties* globally supervene on the *properties of microphysics* is to say that if all the microphysical properties are fixed, everywhere in the world, then there is no more work to do in fixing the chemical properties. They are already fixed. To say that the *social properties* globally supervene on the *individualistic properties*, then, is to say that if all the individualistic properties are fixed, everywhere in the world, then there is no more work to do in fixing the social properties.

But that is just an intuitive characterization. More specifically, supervenience claims are always about how distributions of properties co-vary, or change in sync with one another, across different possible worlds. Here is a common formulation of the global supervenience of a set A of properties on a set B of properties:

> A-properties globally supervene on B-properties if and only if for any worlds w_1 and w_2, if w_1 and w_2 have exactly the same worldwide pattern of distribution of B-properties, then they have exactly the same worldwide pattern of distribution of A-properties.[11]

Applying this to social and individualistic properties, if two worlds have different distributions of social properties, then they also must have different distributions of individualistic properties.[12] Variations in the social properties entail variations in the individualistic properties.

A Diagnostic Tool

Global supervenience can be a good diagnostic tool for assessing whether a thesis like (OI1) is true. Suppose (OI1) were false. Then there would be some social fact, in some world, that was not fully grounded by any set of individualistic facts in that world. The most natural way for that to happen is if there is something *else* that makes that social fact obtain, other than the individualistic facts. So if we change that other thing, the social fact will change, without having changed the individualistic facts. Global supervenience would likely fail.

Conversely, suppose global supervenience fails. That is, suppose there is a difference between the distributions of social properties between two worlds

---

[11] McLaughlin and Bennett 2005. See also Kim 1984.

[12] The different versions of global supervenience correspond to different ways of interpreting "worldwide patterns of distribution." See Bennett 2004a; Shagrir 2002; Sider 1999, 2006.

without there being a difference in the distributions of individualistic properties. How can there be such a difference? The most natural way is if there is at least some social fact, in some world, that is not fully grounded by any set of individualistic facts. Which means (OI1) would be false.

But the connection between (OI1) and global supervenience is not perfect. One can succeed without the other succeeding, and one can fail without the other failing. As I say, global supervenience is good but not infallible evidence. Many philosophers have noticed that the co-variation of social with individualistic facts can never be enough to guarantee that the individualistic facts are the *metaphysical reason* for the social facts obtaining. We can have whatever version of supervenience we want, and it still does not give us grounding.[13] Supervenience is not enough to capture dependence, and we should not expect it to be. The force of a thesis that neuters dualism cannot be just that social facts co-vary in the right way, with respect to individualistic ones. It has to be that they are metaphysically *built* out of the individualistic ones. To take a supervenience relation to be the same thing as a dependence relation is tantamount to confusing the ultrasound with the pregnancy.

A Practical Shortcoming

I want to stress a different shortcoming of supervenience—more practical than principled, but still serious. Supervenience is most commonly understood as a relation between sets of properties, not sets of facts. It is not that supervenience cannot be understood as a relation between sets of facts. The problem, though, is that this conversion is often done wrongly. People often misunderstand the relation between a set of properties and a set of facts. This leads them to mistakenly conclude that a set of facts supervenes on another set of facts when it actually does not.

Consider, for example, the following passage from Jaegwon Kim's seminal paper on supervenience. Kim explains how to turn a claim about the supervenience of facts into a claim about the supervenience of properties.

> A singular fact, I take it, is something of the form a is F, where a is an individual and F a property; and to say that the fact that a is F is a fact of kind P (say, a psychological fact) amounts, arguably, to saying that F is a property of kind P (say, a psychological property). It follows then that for two worlds to be identical in regard to facts of kind P is

---

[13] See McLaughlin 1995; Rosen 2010, 113–14; van Cleve 1990.

for the following to hold: for any property F of kind P and any x, x has F in one world if and only if x has F in the other.[14]

To apply Kim's statement to social facts and social properties: To say that the fact *a is F* is a social fact amounts to saying that F is a social property. And it follows that for two worlds to be identical in regard to their social facts is for the following to hold: for any social property F and any x, x has F in one world if and only if x has F in the other.

This is a mistake. Consider the fact **The sun is warm**. One might think that this is an example of a "solar fact," but that is not so, according to Kim's definition. It is a "temperature fact," since the property F is a temperature property. Likewise, the fact **The freshman class at Tufts is warm** is a temperature fact, on Kim's definition, but not a social fact. In other words, the way Kim translates fact supervenience into property supervenience is to drop the part that is not a property.

As I pointed out in introducing social facts, it is not correct that a fact of the form *a is F* is a social fact just in case F is a social property. It is also social fact if *a* is a social object. Consider two worlds, one in which the freshman class at Tufts is cold, and the other, in which the freshman class at Tufts is warm. These worlds differ in their social facts. But Kim's definition would imply that they do not.

Because supervenience is most perspicuously understood as a relation between property sets, it leads to mistakes when we consider sets of facts. In particular, it leads people to underestimate what social properties need to be included. If supervenience is to address the relation between social and individualistic facts, we need to do one of two things: (1) come up with a form of supervenience that explicitly relates sets of facts, or (2) expand the set of properties so that we are sure that *all* the social facts are taken care of. There are several ways we can appropriately expand the set of social properties. For instance, we can include in the set of social properties all the "identity properties" for all the social objects. That is, properties of the form *being an a*. (For example, the property *being the freshman class at Tufts*.) A different option is to turn the social facts into properties: for every social fact *f*, we could include the property *being such that f obtains*.

This problem is not unsolvable, but it is often overlooked. People take some limited set of properties, apply them to a limited set of objects, and conclude that supervenience works. Supervenience has been misapplied by so many people for so long that I think it is fair to criticize the tool as misleading.

[14] Kim 1984, 169.

Again, this leads us back to preferring talk about facts, and grounding and determination and dependence relations among sets of facts. Supervenience can be an effective diagnostic tool when it is applied properly. But it is a tricky piece of machinery. The current approach tends to leave out literally half of every social fact. This mistake, in my view, is a key reason that 30 years have gone by since supervenience was first applied to ontological individualism without people realizing that ontological individualism is flawed.

9

Against Conjunctivism

Ham sandwiches are not kosher. Ham is made from pigs, and pigs aren't kosher animals. Why are pigs not kosher? Because according to the laws of "kashrut," listed in Leviticus, for a land animal to be kosher, it needs to have a cloven hoof and to chew its cud. "The swine, because it parts the hoof and is cloven-footed, but does not chew the cud, is unclean to you." (Lev. 11:7) These are the grounds for the fact that ham sandwiches are not kosher.

I have sharply distinguished grounds from anchors. The fact **Ham sandwiches are not kosher** is *grounded* by the following four facts: **Ham sandwiches are made of ham**, **Ham is made from pigs**, **Pigs are land animals**, and **Pigs do not chew the cud**. Different facts *anchor* the frame principle expressed in Leviticus 11:7. The anchors might be facts about beliefs in the community, facts about practices over time, or facts about divine commands. The anchors, however, are not among the grounds. The frame principle gives the conditions for grounding the social facts, and the anchors set up these grounding conditions.

Although this distinction seems natural—as it should!—it represents a sharp break from the prevailing orthodoxy. In this chapter, I want to explain and confront the dominant view, which I will call "conjunctivism." This is the view that the grounds for a social fact *include* the anchors, in addition to what I am calling the grounds. Anchors, according to the conjunctivist, are just another kind of ground. Conjunctivism seems attractive, for reasons I will discuss. However, it gets the grounds wrong for many social facts.

The Claim of Conjunctivism

The conjunctivist makes a strong claim about social facts: every social fact has two different kinds of grounds. The grounds for a social fact include those I have been calling grounds, *and also* include the facts I have been calling anchors. Many followers of Searle, for instance, think that the fact **Whitey is a murderer** is grounded by two different facts:

(9.1) ***Whitey killed a person with deliberately premeditated malice aforethought***, and

(9.2) ***People in the United States collectively accept that people who kill with deliberately premeditated malice aforethought count as murderers.***

According to philosophers like Searle, the first fact alone does not suffice to ground the fact that Whitey is a murderer. More generally, they think that a fact *z is Y* is grounded by the conjunction of two facts:

(9.3) *z is X*,

and

(9.4) ***We collectively accept that Xs count as Ys in context C.***

Implicitly, the conjunctivist believes that we are *impotent* to introduce social facts that are grounded nonconjunctively. No social fact, for instance, could be grounded only by (9.1). According to the conjunctivist, the grounds of all social facts conform to a two-part template.

In this chapter, I argue that conjunctivism is implausibly restrictive. It gets the grounding conditions wrong for many social facts.

Still, rejecting conjunctivism is no small matter. A sharp distinction between anchors and grounds involves a major change in how we model possibility. When we anchor a frame principle, we set up the grounding conditions—the full grounding conditions—for a given type of fact. The frame principle gives grounding conditions for *all* possibilities in the frame.[1] And those grounding conditions do not include the anchors themselves. In standard models of possibility, there is just one universe of all the possible worlds. Rejecting conjunctivism, however, suggests that we model possibility differently. A frame should be understood as containing a full universe of possible worlds. When we anchor new frame principles, we move to a different frame: that is, to a different universe of possible worlds.[2]

---

[1] Typically, at least. As I mentioned in chapter 5, I am skeptical about necessitarianism about grounding. This issue, however, plays no role at all in the present chapter. (The only role the rejection of necessitarianism plays in this book is to allow me to distinguish grounding from determining.) All the frame principles I consider are frame-necessary. I am grateful to Dilip Ninan for raising this issue.

[2] To be more precise, this can be modeled using multidimensional modal logics, with multiple universes, or using multiframe logics, with multiple Kripke frames. See, for instance, Grossi 2007; Grossi, Meyer, and Dignum 2005, 2006, 2008. Grossi and collaborators do not discuss anchoring, but provides a series of relevant multiframe logics.

This is a big revision to standard models of possibility, and it is natural—and probably wise—to be skeptical. The case against conjunctivism is convincing, in my view, and rethinking our model of possibility is powerful and clarifying. Still, I expect it to be controversial, and so I have put it at the end of this part of the book. Even readers who favor conjunctivism do not need to stop reading here. None of Part Two turns on the denial of conjunctivism. Everything to follow is compatible with both conjunctivism and its denial.

The conjunctivist, however, should also be aware of another cost to her view. Conjunctivism makes the defense of individualism even harder. In denying conjunctivism, I am doing the individualist a favor. After all, the conjunctivist takes the grounds of a social fact to include both its anchors and its grounds. For the conjunctivist, there is no distinction between ontological individualism and anchor individualism. To be an individualist, then, one cannot just defend one sort of individualism or the other, but instead needs to defend both.

Why Conjunctivism Seems Appealing

Let us suppose that Searle's theory of anchoring dollars is correct. And suppose that the piece of paper I called 'Billy' is a dollar. Isn't it a requirement for Billy to be a dollar that the anchors obtain? After all, consider the following counterfactual: *If we did not collectively accept the constitutive rule, then Billy would be just a piece of paper, not a dollar.*

This counterfactual seems like a slam-dunk argument for conjunctivism. The counterfactual certainly seems to be true. And if so, it seems to contradict the idea that the anchors are not among the grounds. A reasonable way to understand this counterfactual is exactly the opposite of what I have said: the facts about collective acceptance are among the grounding conditions for the fact **Billy is a dollar**. This suggests that the grounding–anchoring model is deficient for capturing how **Billy is a dollar** is grounded. Instead, it suggests that figure 9A is a better picture:

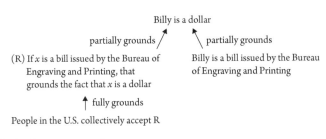

Figure 9A Anchors as grounds

The key difference between figure 9A and the grounding–anchoring model is that in 9A, the frame principle does not just articulate what the grounds are for being a dollar. Instead, the frame principle *itself* is part of the grounds for the fact **Billy is a dollar**. The frame principle, in turn, is grounded by the "collective acceptance" facts. We can still use the term 'anchors' to denote certain grounds of frame principles. But these anchors are just more grounds. Accepting this alternative picture, the thesis of ontological individualism becomes a thesis about both grounds and anchors. On this alternative, when we say that the grounds of social facts must be individualistic, that means *both* the grounds for frame principles *and* the other grounds must be individualistic.

The argument in favor of conjunctivism seems clear enough. To see why it does not work, I first want to look at the counterfactual. At first glance, it seems like decisive evidence, but under scrutiny it turns out not be evidence at all. Then I will consider the version of conjunctivism depicted in figure 9A. That picture turns out to be flawed, and probably cannot be fixed. Subsequently I will return to the point that conjunctivism gets the grounding conditions wrong for many social facts. And finally, having argued against conjunctivism, I discuss why it does not make sense to take ontological individualism to be a claim about anchoring.

The Counterfactual

If we did not collectively accept R, then Billy would be just a piece of paper, not a dollar. This counterfactual seems to be obviously true. And that seems to imply that collective acceptance of the rule is among the conditions for being a dollar.

Counterfactuals like these, however, are slippery pieces of data. In fact, we can interpret this kind of counterfactual in many different ways, some of which are compatible with collapsing the anchors into the grounds, and some of which are compatible with keeping the anchors separate from the grounds. It does not provide evidence one way or the other.[3]

A standard way to interpret the counterfactual is using "possible worlds semantics."[4] On this interpretation, the counterfactual means the following: take a look at the possible world that is nearest to the actual world, but in which we do not collectively accept the constitutive rule R. In that world, Billy is not a dollar, but is merely a piece of paper.

---

[3] I am grateful to Jody Azzouni for extensive comments.
[4] See Lewis 1973; Stalnaker 1968.

On the grounding–anchoring model, however, there is a different natural way to interpret a counterfactual. We not only have possible worlds, but also possible frames.[5] So the antecedent of a counterfactual can shift us in two ways: to the nearest possible world in which the antecedent obtains, within a frame; or to the nearest possible frame in which the antecedent is an anchor, while we remain in the actual world in that frame.[6] The example of money may make this difficult to see, so let us illustrate with a cleaner example.

Consider the fact *x is a senior citizen*. The grounding condition for this fact is *x is at least 65 years old*. This is anchored perhaps by law, by convention, by collective acceptance, or by practices. Now consider the following counterfactuals about Max, a 66-year-old:

(9.5) If Max were 60 years old, he would not be a senior citizen.

(9.6) If we changed the laws, conventions, and practices such that the conditions for being a senior citizen are being 75 years or older, Max would not be a senior citizen.

On the conjunctivist model, both of these counterfactuals have natural interpretations. The antecedent of (9.5) takes us to the nearest world in which Max is 60 years old. In that world, the anchors still obtain, so the conditions for being a senior citizen are the same. And so Max is not a senior citizen, and the counterfactual comes out true. The antecedent of (9.6) takes us to the nearest world where the laws are different. In that world, Max is 66 years old but the second condition for Max being a senior citizen does not obtain. And so Max is not a senior citizen in that world, and the counterfactual comes out true.

However, both of these counterfactuals have natural interpretations on the grounding–anchoring model as well. The antecedent of (9.5) is naturally read as world-shifting. It shifts us to a different world in the frame, that is, the nearest one in which Max is 60 years old. Since it is within the frame, *being a senior citizen* has the same grounding conditions as it actually does, and so Max is not a senior citizen, and the counterfactual comes out true. The antecedent of (9.6) is naturally read as frame-shifting. It shifts us to the nearest frame in which the frame principle is anchored differently. In the actual world with frame principles anchored as in that shifted frame, Max remains 66 years old, but the conditions for being a senior citizen are different. So Max is not a senior

---

[5] For discussion of a closely related point, see Einheuser 2006. Einheuser gives a fragment of a two-dimensional semantics for "counterconventional conditionals." This fascinating paper makes significant progress in distinguishing context-shifting and convention-shifting.

[6] An elaborated multiframe model will need to account for how worlds are identified across different frames.

citizen, and the counterfactual comes out true. Thus neither (9.5) nor (9.6) says anything in favor of either view.

Now consider one more counterfactual, whose reading is not so clear:

(9.7) If there were no body of laws, conventions, and practices, Max would not be a senior citizen.

To my ear, counterfactual (9.7) is ambiguous. On one reading, the grounding conditions for *senior citizen*—*our* grounding conditions—are left the same, and all that has changed is that Max is living in a different environment (one without certain laws, etc.). In that case, Max still satisfies our grounding conditions and is a senior citizen, and the counterfactual is false. On the second reading, we are considering a case in which the grounding conditions for being a senior citizen change, and in that context there is no kind *senior citizen* at all. On that reading, the counterfactual is true.

A reasonable interpretation of the ambiguity of (9.7) is this: counterfactuals involving social facts can be employed either to shift us to other possibilities within a frame, or can mark a shift of frame. In this counterfactual, either shift is about as natural as the other. The reason (9.5) and (9.6) are *un*ambiguous is that it is clear what the antecedent does: the first does not shift the frame, and the second does.

But this is not the only reasonable interpretation. Counterfactuals can be modeled in many ways, and the case I have described can just as easily be seen as a shift in the set of relevant possibilities as it can a shift between frames.[7] The fact that (9.7) has two readings, in other words, is compatible with both conjunctivism and its denial.

Altogether, the point is not that the counterfactual is evidence in favor of the separation of anchors from grounds. Rather, it is that the counterfactual is not evidence one way or the other. It is compatible with both views, and cannot be used to decide between them.

A Dilemma for Constitutive Rules

Thus the counterfactual does not favor either conjunctivism or its denial. Why reject conjunctivism? Let us start with the version I depicted in figure 9A. This seems like a natural picture for Searle's theory.

---

[7] For discussion of counterfactuals and modal bases, Portner 2009 and Kratzer 2012 are particularly useful.

In figure 9A, one of the two grounds of **Billy is a dollar** is constitutive rule R. That rule says that if *x* is a bill issued by the Bureau of Engraving and Printing, that *fully grounds* the fact that *x* is a dollar. Unfortunately, the diagram contradicts that rule. Satisfying the rule's conditions only *partially* grounds the fact that *x* is a dollar. The other partial ground is the rule itself. So 9A cannot be the right picture.

Perhaps the answer is to change the diagram, leaving the rule out. This would give us a different figure (figure 9B).

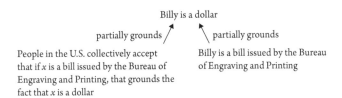

Figure 9B The constitutive rule left out

But this move does nothing to fix the problem. In this diagram, we take only the anchors, not rule R itself, to be among the grounds of **Billy is a dollar**. Yet even if we do this, the rule *still* shows up in 9B: it is what is collectively accepted. So even if we move to the new figure, we still need to fix the rule.[8] If we do not, then the people in the United States are collectively accepting something false. (Again, according to the diagram, *x* **is a bill printed by the Bureau of Engraving and Printing** only *partially* grounds *x* **is a dollar**.)

Unfortunately, the rule cannot be fixed. We cannot revise it, so that it gives us the full grounding conditions. If we include the full grounding conditions in the constitutive rule—that is, if we change it so that we collectively accept the correct conditions for being a dollar—then we fall into an infinite regress.[9] To see this, let us try to revise the rule, so that it is both the thing we collectively accept, and also the correct rule for being a dollar.

There is not just one condition for something to be a dollar, but two. Therefore, rule R cannot be what we collectively accept. Instead, we need to replace R with R-partial:

(R-partial) If *x* is a bill issued by the Bureau of Engraving and Printing, that *partially* grounds the fact that *x* is a dollar.

---

[8] This same point applies to nearly all versions of the Standard Model, not just Searle's. It applies to views in which attitudes are taken toward the rule itself. Hume's theory is not included, but Hart's is, as are other prominent theories in the recent literature, such as Raimo Tuomela's "collective acceptance thesis" (Tuomela 2007, 187) and Frank Hindriks's revised constitutive rules (Hindriks 2009).

[9] Searle 1995, 52 discusses another potential regress. The present difficulty is not the same.

R-partial does not give the full grounds for *x is a dollar*: it gives only partial grounds. If we are to collectively accept a rule that gives us the full grounding conditions for *x is a dollar*, we need to collectively accept a rule that has *conjunctive* grounding conditions:

(R*) If x is a bill issued by the Bureau of Engraving and Printing *and* we collectively accept R-partial, then together those ground the fact that x is a dollar.

We have to accept R*, if we take ourselves to accept the correct conditions for being a dollar. Because only R* gives the conjunctive grounding conditions.

However, that is still not enough. After all, now we are collectively accepting R*, and *that* acceptance is also part of the grounding conditions for being a dollar. Therefore, R* is not correct: it only gives the *partial* grounding conditions for *x is a dollar*. To fix this, we need to replace R* with R*-partial—that is, substituting 'partially ground' in the place of 'ground':

(R*-partial) If x is a bill issued by the Bureau of Engraving and Printing *and* we collectively accept R-partial, then together those *partially* ground the fact that x is a dollar.

And then the rule giving the full grounding conditions involves a bigger conjunction:

(R**) If x is a bill issued by the Bureau of Engraving and Printing *and* we collectively accept R-partial *and* we collectively accept R*-partial, then together those ground the fact that x is a dollar.

It is R** that gives the conditions for being a dollar, not R*. But once again we find ourselves in the same situation. If we want to insist that we accept the rule for what it takes to be a dollar, then we also have to be *accepting* R**. So R** needs to be replaced with R**-partial, and accepting *that* has to be part of the conditions for being a dollar. Which takes us to R*** and R***-partial. And then to R**** and R****-partial. And so, ad infinitum.[10]

To break the regress, we have two options. We can drop the claim that we collectively accept the conditions for being a dollar, or else we can drop conjunctivism. To put it differently: the conjunctivist has a Hobson's choice. Either the rule we accept is false, or else it is limited to capturing a partial set of

---

[10] One might take a self-referential approach, along the lines of Peacocke 2005, but then it becomes implausible that we collectively accept the complex rules that would require.

conditions for being a dollar. For the conjunctivist, the constitutive rule cannot accomplish what the "Standard Model" takes it to: namely, to give the conditions for a social fact to obtain.

I am no proponent of Searle's collective acceptance theory, nor of Hart's or any other theory following the Standard Model. Nonetheless, I doubt that the views of Hart, Searle, and others are so easily self-defeating.[11] I suggest we take Searle at his word: what we collectively accept is the rule for being a dollar, or a boundary, or a president. But that means that the anchors are not among the grounds: the acceptance of the rule is not itself among the conditions for something to be a member of its kind.

Conjunctivism Gets the Grounding Conditions Wrong

The argument in the previous section applies specifically to Searle's and others' "Standard Model" approaches. Given that argument, it is always an option to abandon the idea that constitutive rules do what the Standard Model claims: that we collectively accept some proposition that gives the grounding conditions for social facts of some kind. If we abandon that idea of constitutive rules, then conjunctivism could still be viable. The real test, however, is which approach gets the grounding conditions for social facts right. And this is the fundamental reason for rejecting conjunctivism. To include anchors as grounds would get the grounding conditions for many social facts wrong.

As I discussed in chapter 5, social kinds and facts are a sort of "universal tool": even though we may anchor their frame principles idiosyncratically, they can be grounded in any situation whatever. We can look back at ancient societies, and evaluate whether there are classes or castes, aristocrats or serfs. We might look for baristas in the Ottoman Empire or in seventeenth-century England, and variable annuities among the ancient Egyptians. We might find that the Egyptians do not have variable annuities, but only proto-annuities. Or we might find that there is, in their context, an entity satisfying the conditions for *being a variable annuity*. Social kinds and social facts are applicable across a universe of different situations. In assessing the grounding of social facts across possibilities, we take the grounding conditions to be fixed.

---

[11] The same argument holds for other leading theories taking a Searle-like approach. For instance, Tuomela's and Hindriks's approaches are each consistent only if the collective acceptance facts are not among the grounding conditions.

To decide whether conjunctivism is right or wrong, we need to assess its implications across other possibilities. That is, to think about whether a given social fact can obtain in possibilities where the anchors for its frame principle do not obtain. Complex cases like money are hard to resolve. But for many examples, it should be relatively uncontroversial that conjunctivism is too rigid. Certain social kinds, for instance, we explicitly anchor to apply *across all situations*, even retroactively.

Consider, for instance, the conditions for *being a war criminal*. One is a war criminal if one has committed or conspired to commit any of a long list of crimes in association with armed conflict. We can sensibly ask whether Caligula was a war criminal, or whether Genghis Khan was, having killed over a million inhabitants of a single city. We can also consider a possibility in which some virtuous person instead committed terrible crimes, and sensibly ask whether that person would be a war criminal. It does not matter whether, in that possibility, there is an International Criminal Court. What matters is only whether the person satisfies the conditions we have anchored.

To collapse anchors into grounds would improperly restrict the social facts to ones having two-part grounding conditions. It would restrict the social facts to ones whose grounding conditions not only include the ones we want them to have, but also all the anchors involved in putting the conditions in place. This is simply not how we use social facts. They can have simple grounding conditions. And when we assess them across other times and possibilities, we do not deny that they obtain merely because the anchoring facts do not obtain at those times and possibilities.

Making Ontological Individualism Sensible

In the last chapter's examination of ontological individualism, I developed it as a thesis about full dependence, taken in a generalized way:

(OI1) For any possible world w, and any social fact f at w, there is some subset X of N (the set of possible individualistic facts), such that X grounds f at w.

In that analysis, however, I avoided discussion of frames and anchoring. In particular, I said nothing about whether "all worlds" are the ones within a frame or across frames, nor did I separate this from a thesis about anchoring.

There is, however, a clear way to understand this, which is both consistent with and charitable toward the original ontological individualism tradition. The idea is to take the thesis to be about just *one* frame, usually our current frame. Anchor individualism is a thesis about how anchoring in general can possibly work. Ontological individualism is a thesis about how social facts in our frame can possibly be grounded, given that the anchors are what they are. I will briefly explain why this is the best alternative.

Ontological Individualism Is Not Anchor Individualism

Some advocates of the Standard Model are tempted to regard ontological individualism as a thesis about what I am calling "anchors." That, however, is not a viable option.

The reason is this: a social fact's grounds have at least *something* to do with making it obtain. Regardless of how we anchor the grounding conditions for *x is the Senate*, or for *x is the freshman class at Tufts*, ontological individualism holds that the social facts depend on the people in those groups. A fact like **The freshman class at Tufts is studious** is grounded—at least in part, if not fully—by facts about the individual freshmen. That is what ontological individualism is a thesis about—about grounding, not about what anchors the conditions for *x is the freshman class at Tufts*. Or maybe it is about both the grounds and the anchors. But it is not about the anchors alone.

If we take ontological individualism to be just a thesis about anchoring, and to say nothing about grounding, then supervenience just about *always* fails. It is too easy to set up situations in which two social domains are exactly alike in terms of the individuals and the individual relations composing them, and yet do not share the same social properties at all.[12] Ontological individualism is at the very least a thesis about the grounds of social facts, that is, about what sorts of other facts can possibly ground social facts. It is at least that.

No Individualist Should Prefer a Thesis about both Grounds and Anchors

Is ontological individualism then a thesis about *both* grounds and anchors? That is, about how frame principles can possibly be anchored *and* about how social facts can possibly be grounded in any frame? This is the position the conjunctivist must take. I have argued that it is an error to collapse anchors into

---

[12] Here is a more technical illustration of this. Take some property Z that is neither individualistic nor social. Anchor a new frame principle as Searle might: we collectively accept that if an object has Z, it has status Y. Now consider two different worlds. In the first, there are people, but no objects having property Z. The second is individualistically indiscernible, but that there is also one additional object that has property Z. In the second domain, there is an object having status Y. In the first domain, there is not. So supervenience of the social on the individualistic fails.

grounds. I want to point out, however, that the proponent of individualism has an even greater stake than I do in keeping them separate.

Consider the anchors and the grounds for a social fact like **Billy is a dollar**. A theory like Searle's takes the anchors to be individualistic—individual "we-attitudes" toward the constitutive rule.[13] But Searle has no stake in taking the grounding conditions to be individualistic. The "X term" of his constitutive rule can involve facts about pieces of paper and lines of stones—things which are neither social nor individualistic. The very idea of his theory is that our attitudes project features onto the brute world of nonpersons.

In general, the anchor individualist should want to keep the grounds of properties like *being a dollar* separate from the individualistic anchors. The anchor individualist should have no interest in being an individualist about grounds.

Conversely, the individualist about grounds will take issue with anchor individualism. There are various strategies the individualist about grounds could pursue, in arguing for her claim. She might tightly circumscribe which facts count as the social ones. She might, for instance, argue for individualism about the grounding of certain macroeconomic facts, but not other facts. Or alternatively, she might add epicycles to the set of facts that count as individualistic. These are the sorts of strategies that theorists working in the tradition of Watkins, Lukes, Kincaid, Pettit, and others pursue.

But whichever moves are made to preserve the thesis that the *grounds* of social facts are individualistic, these are entirely different from those made by the anchor individualist. The individualist about grounds gains nothing by taking a position on how anchoring occurs. The collapse of anchors into grounds only makes her task harder.

To many people, both ontological individualism and anchor individualism are appealing for the same reason. Both seem to deflate worries about dualism with regard to the social world. The social world is just *us*, both theses hold. However, the two theses deflate these worries in conflicting ways. For an ontological individualist, the prototypical example of a social fact is one about a group, like a court or legislature, which is composed of individual people. The ontological individualist typically regards social facts as emerging from interactions among individual people, in combination with one another. For an anchor individualist, in contrast, the prototypical example is a fact about dollar bills, unkosher animals, or boundaries made of lines of stones. It is not just that the strategies are distinct. It is that they are at odds with one another.

---

[13] Recall that in Searle's view, a "we-attitude" is not a group-dependent attitude, but simply a different sort of intentional state that an individual can be in, different from that person's "I-attitudes" (Searle 2010, 42–50).

An Olive Branch

Having argued for the model, and for the sharp separation of anchors from grounds, I want to extend an olive branch to skeptics, agnostics, and dissenters.

The reason I argue for the sharp separation of anchors from grounds is that I think it makes for a better model, and that it illuminates the project of social metaphysics. It also makes it easier to see the flaws with anthropocentrism about the social world. One reason these flaws have been so hard to see is that the two different kinds of individualism have been confused with one another. And as I have said, however, separating the two theses does a favor for the individualist. It keeps individualism at least a little plausible, and makes us do some work to show why it fails.

The discussion of groups in coming chapters does not at all depend on the sharp distinction. I have found the grounding–anchoring model to be more powerful, and to produce more insights, so I do not see the point in hedging my bets. Still, all the rest of the points of the book remain intact, even for the conjunctivist. Conjunctivism does not compromise the force of the arguments before or after this chapter. If it turns out to be correct, that means only that the grounding inquiry I pursue is not an account of the full grounds, but only of partial grounds. And it means that the anchoring inquiry must be completed as well in order to get an account of the full grounds of social facts. Thus for the conjunctivist, the following chapters on the grounding of facts about groups will be part of the story, but not the whole one. Nonetheless, the same failures of individualism and consequences for modeling still apply.

Grounding, Anchoring, and the Social Sciences

How is the grounding and anchoring of social facts pertinent to social science? How social facts are grounded matters for how we build models of them. And in a different way, how frame principles are anchored also matters for modeling. Of the two, grounding is the more directly relevant. As a result, the truth or falsity of ontological individualism is a more urgent matter for social science than is the truth or falsity of anchor individualism.

Suppose, for instance, there are too many first-degree murders in our community, and we want to figure out how to reduce their numbers. Imagine a politician coming to the stump and proposing, "Under my administration, each of us will accept different conditions for being a first-degree murderer in our community. Here is what I propose: we collectively accept that only people over 80 years old are first-degree murderers. Then the rate would plummet!" Good idea, but not exactly the point. If we are interested in policy changes to reduce the incidence of first-degree murder, we leave the grounding conditions fixed. If we want to model this incidence, we could model the factors that affect the grounds for different incidences of first-degree murder. We might see what factors have a causal effect on killings, such as accessibility of guns. We might see what factors have an effect on premeditation, such as crime shows on TV. To reduce the incidence of first-degree murder, however, we would not model changes in the anchors of the frame principle recorded in MGL c. 265 §1. In trying to change the incidence of first-degree murder, we are not normally interested in modeling what would change the conditions for having the status of a first-degree murderer. Instead, we take the anchors to be fixed, and model what affects the grounds.

That is not to say that anchoring is always irrelevant to modeling in the social sciences. In some contexts, the anchors change rapidly, so we may be interested in assessing what affects them. On Wall Street, for instance, there is rapid innovation in financial instruments. New types of derivative instruments are continually created, not just simple options and swaps, but things like quantos, basket options, diff swaps, and so on. Sometimes they are created by explicit contracts, and sometimes by the practices of financial traders. Effectively, these people are anchoring new social kinds, setting up entities with different grounding conditions than there were before. Once such a kind is set up—that is, once various frame principles for facts about such kinds are anchored—then facts about particular instances of these kinds can be grounded. If we are interested in modeling financial markets, we may just want to take the set of financial kinds fixed, anchored as they are, and see how changes in the world affect facts about them. But instead, we might want to model the dynamics of anchoring in these markets, to see how innovation can be affected.

Still, when we talk about the supervenience of social facts on facts about individuals, we are talking about grounding, not anchoring. When we talk about the emergence of social facts from facts about individuals, we are mostly talking about grounding, not anchoring. Most importantly, when we build models in the social sciences, we are mostly interested in the grounds of a set of possible social facts, and the causal factors that can affect those grounds. Typically, we take the anchors, and hence the frame principles, and hence the grounding conditions for social facts, to be fixed. And we model how changes in the world affect the grounds of social facts, thereby changing which social facts obtain.

PART TWO

GROUPS AND THE FAILURE OF INDIVIDUALISM

This part of the book turns to groups of people, and in particular to the *grounding* of facts about groups. One aim of this part is to put ontological individualism to bed. I already hinted at its flaws back in chapter 3. But here I move past mere hints, and dispense with even the most modest version of the thesis.

A more important aim, however, is to build, not just to criticize. To advance the "grounding inquiry," we do not want just to argue the negative point that social facts are not as anthropocentric as people have assumed. We want, instead, to dive into the work of figuring out frame principles for real social facts.

So why groups? If we want to refute ontological individualism, there are surely easier cases. Ontological individualism, after all, is a *universal* claim. It holds that *every* social fact—without exception—fully depends on some set of individualistic facts. To refute a claim like that, we need to find just one single social fact that is not fully grounded by facts about individuals. Maybe we can show this for a fact about Starbucks Corporation or Tufts University. Or a fact about money, or credit default swaps.

Starbucks, of course, was the example I used in chapter 3. It is folly, I suggested, to expect that facts about Starbucks supervene on facts about individual people, even understanding "facts about individual people" very broadly. That would be like expecting facts about human anatomy to supervene on facts about cells. Intuitively, this is an easy case: it is reasonably clear that Starbucks is a vast and heterogeneous stew of constituents. So why not just examine facts about Starbucks, or Tufts, or money, or credit default swaps? Why consider facts about groups? If the aim is to challenge anthropocentrism, groups seem like the very *worst* example we could choose. After all, groups of people are made of nothing but people. Surely, facts about groups are more likely to be individualistically grounded than are facts about Starbucks.

This, of course, is exactly why I am focusing on groups. By showing that even basic facts about groups of people are not individualistically grounded, the case is even stronger for all the other social facts. If we only considered facts about Starbucks, we might wonder how generally applicable the failure of ontological individualism is. But once we recognize its failure for facts about groups, we will have no such worries.

And there is another reason to focus on groups. Groups are easier. Things like corporations and money are devilishly complicated, much more so than Searle and others suggest. They are anchored by elaborate social structures, and their grounding conditions alone would fill books. Choosing one of those as an example would mean that we could not make any real progress on the second aim—that is, on advancing the grounding inquiry. In this regard, the simplicity of groups is a distinct advantage.

This part consists of eight chapters. Chapter 10 addresses the nature of groups, and the notion of "constitution" as it applies to groups. In chapter 11, I introduce some simple facts about one particular group, and work through a handful of frame principles for some of those facts. In chapter 12, I discuss "criteria of identity" and examine how we can identify groups over time—even at times when they have no members at all. And in chapter 13 I use that machinery to work through a couple of more complicated frame principles. These are frame principles for facts about groups of a given *kind*, rather than facts about a *particular* group. That chapter is as deep as I get into the weeds of frame principles.

In chapters 14, 15, and 16, I examine group attitudes—in particular, group intentions. Chapter 14 introduces the problem, and discusses various patterns of grounding for social facts. Chapter 15 argues that group actions are not fully grounded by the actions of group members. And chapter 16 extends the point to group intention: the intentions of a group are not fully grounded by the attitudes of its members. This contradicts a basic assumption in the literature about group intentions, and group attitudes in general.

In chapters 17 and 18, I return to the question of the nature of groups. I address two alternative approaches to groups, the "social integrate" models and the "status" models. I show that the narrow conceptions that prevail in the literature are merely special cases of the broader model I present in this part of the book. Finally, I discuss directions for future inquiry, and draw connections between the failure of individualism and modeling in the social sciences.

10

Groups and Constitution

What is a group? I will take a broad view. A group is a thing constituted by and only by individual people. Many of the entities we investigate in the social sciences are groups. The Supreme Court is a group, as is the United States Senate. The president's cabinet is a group, as is his family. The faculty of Tufts is a group, the American people is a group. So, too, is the workforce of Starbucks and the American tax bureaucracy. The proletariat is a group, the bourgeoisie is a group. The wealthy make up a group, as do the poor, senior citizens, and the mob storming down the street. Still, far from everything in the social world is a group. Money is not a group, stock options are not groups, nor are restaurants, borders, or promises. Corporations and universities are not groups, though their boards, workforces, faculties, and student bodies are.

As I will discuss, this is broader and more generic than many treatments of groups in the literature. And, my characterization is not as innocuous as it might look: it is simple, but involves the technical term 'is constituted by'. *Constitution*, like grounding and some other relations I have discussed, is a new tool that has recently received a lot of attention in metaphysics.

The aim of this chapter is to explain the broad characterization of groups, and especially to make sense of constitution as it applies to groups. I do not yet explore how facts about groups are grounded. That is the topic of chapter 11. To begin, I will lay out some of the paradigms of groups in the literature. Then I will turn to constitution.

Four Paradigms of Groups

In the social sciences, we find several different paradigms of groups, with theorists often focusing on only one or two. One paradigm that periodically captures the popular imagination is the mob or crowd. Worries about the crowd rose through the nineteenth century, at a time when mass movements exercised great political power.[1] In 1895, Gustave Le Bon wrote a bestseller titled *The Crowd: A Study of the Popular Mind*, in which he argued that crowds represent an atavistic, wild, and premodern state of man. Le Bon's book is full of unhinged pronouncements, like this one:

> Among the special characteristics of crowds there are several—such as impulsiveness, irritability, incapacity to reason, the absence of judgment and of the critical spirit, the exaggeration of the sentiments, and others besides—which are almost always observed in beings belonging to inferior forms of evolution—in women, savages, and children, for instance.[2]

Despite claims like these, Le Bon's paradigm of a group as a crowd or mob has remained influential. Indeed, nearly every twentieth-century political system—democracy, communism, and fascism included—has been analyzed as a crowd movement at one point or another. Here, for instance, is Elias Canetti on the "spontaneous crowd," in his Nobel prize-winning commentary on fascism, *Crowds and Power*:

> As soon as it exists at all, it wants to consist of more people: the urge to grow is the first and supreme attribute of the crowd. It wants to seize everything within reach; anything shaped like a human being can join it.[3]

Even today, the influence of this paradigm continues. In recent years, a number of popular books have come out that invoke it, such as *The Wisdom of Crowds* and *Crowdsourcing: Why the Power of the Crowd is Driving the Future of Business*.[4]

A different paradigm of groups is the large and heterogeneous mass, such as an economic class or national population—a group of people somehow united such that they can be regarded as a single entity acting in systematic ways. These are the kinds of things that Alexis de Tocqueville has in mind in *Democracy in America*, when he speaks of the American people as a whole, as well as of other large-scale associations such as the class of lawyers in America.[5] Karl Marx uses

---

[1] Schnapp and Tiews 2006.
[2] Le Bon 1895, 16.
[3] Canetti 1962, 16.
[4] Howe 2008; Surowiecki 2005.
[5] de Tocqueville (1889) 2003.

a similar conception of groups in his theories of social change, identifying a few large economic classes as the key explanatory units for social theory:

> There are three great social groups, whose members, the individuals forming them, live on wages, profit, and ground respectively, on the realization of their labor power, their capital, and their landed property.[6]

If a theorist's aim is to generalize about large-scale social phenomena, it is natural to think of groups like these as basic social units. A science like astronomy or meteorology generalizes about large-scale, heterogeneous objects such as planets or clouds. Similarly, on this conception the social sciences generalize about groups that are not formally or intentionally structured. The idea is typically that these amorphous groups conform to social laws, much as the planets conform to physical laws.[7]

A third paradigm of groups relies on formal organizations. This paradigm emerged as social theorists started to think about the modern world developing regimented structures of social interaction. Max Weber, instead of thinking of groups as large-scale entities subject to social forces, emphasizes their authority and management structures. Modernity, on Weber's view, is characterized by the development of the bureaucratic organization:

> Normally, the very large modern capitalist enterprises are themselves unequalled models of strict bureaucratic organization. Business management throughout rests on increasing precision, steadiness, and, above all, speed of operations.[8]

Finally, a fourth paradigm has become popular in recent social theory: groups as small networks of a few individuals, interacting with one another. Common examples of this paradigm are people negotiating with one another, playing a game against one another, or else engaged in some collaborative activity with one another, such as walking together, painting a house together, or performing a jazz improvisation together. This paradigm is often employed in game theory, as well as in theories of the intentions and actions of groups, such as in the work of Margaret Gilbert, Michael Bratman, Philip Pettit, and Christian List.[9]

---

[6] Marx and Engels (1975) 1998, 871.
[7] de Tocqueville (1889) 2003, chap. 12; Marx 1867, §3, Part C.
[8] Weber 1978, 974.
[9] Bratman 1993; Gilbert 1989; List and Pettit 2006, 2011; Pettit 2003.

Margaret Gilbert in particular pioneered the use of a small network as a paradigm for the social group. In her 1989 book *On Social Facts* and her 1990 paper "Walking Together," Gilbert argued that "we can discover the nature of social groups in general by investigating such small-scale temporary phenomena as going for a walk together."[10] Compared to the other paradigms, this one may seem strange. Since so many groups are large, it may be hard to imagine that a network of a few people can be informative for groups in general. Gilbert says as much: "The idea is attractive insofar as it should be relatively easy to understand what it is to go for a walk with another person. It may also seem somewhat farfetched."[11]

As we will see, these impressions are right: the idea is attractive, simple, and useful in certain contexts. But it is also farfetched. For understanding social groups in general, this paradigm is as misleading as it is helpful. Most social groups are not small-scale, not particularly temporary, and not nearly as isolated or collaborative as a couple going for a walk. I discuss this approach and its shortcomings in detail in later chapters.

For now, however, I only want to draw attention to these very different notions of groups employed in social theory. We should be cautious about starting our inquiry with too many restrictions. At least to begin, it pays to be ecumenical about which sorts of groups figure into the social sciences.

On the other hand, we cannot be entirely noncommittal about the characteristics of groups. If we are to model groups in the sciences, or even work with facts about groups at all, we need to take at least something of a stand on how to track groups over time and possibilities.

Groups and Sets

When we model objects in the sciences, we track them over changes in circumstances. We may be interested in starting with the way an object actually is, and in tracking how it changes over time. Or we may be interested in how an object *might* change, were we to change certain of its features. Or in how an object might change, were we to change its circumstances. A key goal of modeling is to investigate changing properties of objects across time and across possibilities.

When we aim to track a given object, we aim to track *it*—that very thing—as it changes across time and possibilities. Suppose, for instance, we want to track the actual changes of a few ant colonies, as they compete for territory. To

---

[10] Gilbert 1990, 2.
[11] Gilbert 1990, 2.

see how they accrue or lose territory over time, we need to keep track of each colony. We need to pay attention to which is which, and be sure not to mix them up with one another. Likewise, if we want to model the *possible* changes in territorial dominance, we need to track the colonies over various possibilities, keeping track of which is which as the world might change. If we are careless about keeping track of the particular colonies, we will not be able to model how their territorial changes vary over situations.

The same holds true in the social sciences. If we want to model the changing facts about the Supreme Court, a social class, or a legislature, we need to be able to track those objects over time and across possibilities. To do so, we need to identify groups over time and possibilities, and distinguish different groups from one another.

In chapter 5, I distinguished the following two facts:

(5.3) *Bob, Jane, Tim, Joe, Linda, ... and Max ran down Howe Street.*

(5.4) *The mob ran down Howe Street.*

I pointed out that the propositions corresponding to (5.3) and (5.4) have different constituents. Bob, for instance, is a constituent of the first proposition, but not the second. The obvious difference between these two propositions is that the first one is about many things running down Howe Street, and the second one is about one thing running down Howe Street. To fix this, some people have proposed that we understand a group, like a mob, to be a *set*, in the mathematical sense. The fact **The mob ran down Howe Street**, on this proposal, is equivalent to:

(10.1) *The set {Bob, Jane, Tim, Joe, Linda, . . ., Max} ran down Howe Street.*

This simple proposal, however, does not work. One problem, of course, is that it sounds strange to say that a mathematical object ran down the street. But there is a more fatal problem: the members of groups are replaceable, but the elements of sets are not.

Turnover in membership is a characteristic of many groups. In the logic class I teach, students occasionally drop out, and new ones enroll. But even as the students change, the class persists. Because of this, I am able to track features of the class despite changes in enrollment. If, for instance, I wanted to track the attrition levels over the semester—perhaps correlating attrition with the difficulty of the tests—I would track the size of the class as its membership changes. The membership of a group can even completely turn over many times, even as the group persists.

Sometimes, I might want to track the fixed set of students who made up the class at some point in time. I might, for instance, want to track the people who made it to my final exam, as they graduate and take jobs. Still, in a typical group the members can change: they are not essential to it. Any adequate model of a group will have to make sense of this.

The elements of a set, on the other hand, are fixed. Two sets with different elements are different sets, and if some object is actually a member of a given set, then it is necessarily a member of that set. So the mob cannot be the same as {Bob, Jane, Tim, Joe, Linda, . . ., Max}.

If the mob is not identical to that set, does that mean it is not identical to *any* set? In the last few years, some philosophers have described more complex mathematical objects—complex sets and other structures—as analyses of what groups are.[12] For my part, I am reluctant to *identify* groups with any mathematical structure at all, even though we can surely use such structures to *model* groups. And whether groups are aggregates, collections, or pluralities is even harder to assess, because there is no agreement about what those entities are either. Altogether, it is not clear how helpful it is to try to replace groups with some other object—mathematical or otherwise—that seems more respectable. But whatever the case, the mob is surely not the set {Bob, Jane, Tim, Joe, Linda, . . ., Max}, because of the replaceability of members of the group.

Groups and Ordinary Objects

A different idea has recently emerged in the literature: building on the similarities between groups and ordinary material objects, like tables and chairs. Metaphysicians have recently developed some tools for understanding ordinary objects—in particular, the notion of *constitution*—which are promising for bringing out key features of groups.

In certain ways, groups differ from ordinary material objects. Most people, for instance, think that ordinary objects need to be clustered in one region of space and time. It may not be necessary for all the parts and subparts of an object to be strictly contiguous. For instance, two adjacent links of a chain are both part of the chain. But even if the adjacent links are not touching one another, the chain may remain intact. Still, there seem to be limits to how

---

[12] Effingham 2010 proposes that groups are more complex sets, ones that associate sets of individuals with particular times and worlds. Ritchie 2013 proposes that groups are a certain kind of mathematical graph. Others take groups to be aggregates, or collections, or pluralities.

dispersed ordinary objects can be.[13] Groups are different. Contiguity is typically irrelevant to groups. To be a member of the Supreme Court, for instance, you do not have to be located near the other members. (Some groups do resemble material objects even in this way: to be a member of the mob raging down Howe Street, you have to be there.)

Nevertheless, other important characteristics are shared by groups and ordinary material objects. Like groups, many ordinary objects can persist even through changes in their composition. Parts of a ship, for example, can be replaced without destroying the ship, and parts of a human body can be replaced without destroying the body.

An even more interesting characteristic of groups is that two distinct groups can have exactly the same members at the same time—that is, two distinct groups can *coincide*. As of 2014, for instance, the board of the Massachusetts Department of Transportation (MassDOT) has the same membership as the board of the Massachusetts Bay Transportation Authority (MBTA). Despite having the same membership, the MassDOT board is not the MBTA board. Each month, the MassDOT meeting is convened, the members conduct the MassDOT business, and then the meeting is closed. Immediately following that, the MBTA meeting is convened and they conduct the MBTA business. Sometimes the business of the two boards is the same, so they have to discuss the same issues twice.[14]

A flood of work in the recent metaphysics literature has dealt with exactly this characteristic—not for groups, but for ordinary material objects. In the next few pages, I discuss the notion of constitution for ordinary objects and its application to groups. I point out the shortcomings of a recent proposal about this application. I develop an alternative analysis of constitution in terms of grounding, and then round out the conception of groups as constituted by people.

Coincidence and Constitution

The idea that two distinct groups can have the same members is not particularly shocking. More surprising is the claim that two distinct material objects can be in the same place at the same time. The claim and the puzzles it raises

---

[13] Lewis 1986, 211, on the other hand, famously argues for a very liberal understanding of material objects. See also Quine 1960, 170; Quine 1981, 10; Heller 1990, 10; and Sider 2001, 7.

[14] See http://www.mbta.com/about_the_mbta/board_meetings/. Margaret Gilbert discusses similar examples in Gilbert 1989, 220–21, and Gilbert 1996, 199. Also see Uzquiano 2004.

are usually described in terms of the relation between a statue and the lump of clay of which it is made. Is the statue identical to the lump of clay?

There is a simple argument that the statue and the lump of clay must be distinct objects. Suppose the statue were to be crushed. Then it would be destroyed. But crushing a lump of clay does not destroy that lump. The statue is destroyed, but the lump lives on. However, it cannot be that a single object is both destroyed and not destroyed. If the statue and the clay have different properties, then they must be different objects. And they do: the clay has the property *would survive a crushing*, while the statue has the property *would be destroyed by a crushing*.[15]

Five alternative positions have emerged in the literature, in response to this observation. (1) *The coincidence view*: The statue and the clay are distinct objects. These objects are not identical to one another, but they do coincide. That is, they occupy the same place at the same time. (2) *The identity view*: Despite the appearance of different persistence conditions, there is just one object. The statue and the lump are identical. Proponents of the identity view have the burden of showing how the very same thing can have different persistence conditions, which seems to violate a basic principle about the nature of identity.[16] Or else they need to deny that the lump can survive being crushed, or that the statue cannot.[17] (3) *The lump view*: Others, more radically, deny the existence of statues altogether, and insist that of the two, only the lump exists. This view has the disadvantage that it denies the existence of statues, but then does not deny the existence of some ordinary middle-sized objects, such as lumps. Why should statues fail to exist, if lumps exist? (4) *The nihilist view*: Others take an even more radical position, denying the existence of *all* ordinary objects, including lumps. They hold that ordinary objects are an illusion—and perhaps also that the only *real* objects are those of fundamental microphysics.[18] This avoids the uncomfortable claim that some middle-sized objects exist while others do not, but it is radically revisionist about the nature of objects altogether. (5) *The dominant kind view*: Still others argue that there is just one object at a given time, the "dominant" object. At the outset, there is a lump of clay and no statue, then when the statue is formed there is a statue but no lump, and when it is crushed there is again a lump but no statue.[19] This

---

[15] This example is due to Gibbard 1975. See also Baker 1997, 2000; Fine 2003; Johnston 1992; Thomson 1983, 1998; Wiggins 1968, 2001; Yablo 1987.

[16] The principle is Leibniz's Law. It says that, necessarily, for all properties P, and for all objects x and y, $x = y$ implies that if x has P then y has P.

[17] Gibbard 1975; Heller 1990; Lewis 1986; Lowe and Noonan 1988; Noonan 1993; Robinson 1982; Wasserman 2002, 2004.

[18] See Dorr and Rosen 2002, Merricks 2001, Unger 1979, and van Inwagen 1990 for variations on this view.

[19] Burke 1994. See also Baker 1997 and Rea 2000 for related discussion.

position, happily, admits the existence of both lumps and statues, and also preserves the principle of one object in a given place at a given time. But it has lumps flit in and out of existence in a weird way, depending on what they happen to make.

It is tempting to regard this as a pseudoproblem. What difference does it make to sort out our intuitions about the persistence conditions for statues or lumps? Or about two objects occupying the same place at the same time?

Yet, here again, how we treat this makes a practical difference. If we were to write a computer program to track the changing properties of the statue and the lump, should we track them as one thing, or as two? We cannot just conflate the statue and the lump, overlooking their differences. Conflating these can trip up a model, especially as we track these things over time and circumstances. After all, if I want to model the situations under which the statue can be destroyed—perhaps for insurance purposes—it would be foolish to track a lump that is not destroyed even when it is crushed into a ball.[20]

Going with the Coincidence View

An increasing number of theorists endorse view (1), the coincidence view. No doubt, it is peculiar to believe that multiple objects can spatially coincide, that is, can occupy the same space at the same time. But all the available positions come with counterintuitive commitments, and the unattractiveness of the alternatives leads many people to concede that the statue and the lump are distinct and yet coinciding objects.[21]

With this, however, comes an explanatory burden. Despite not being identical, the statue and the lump are close relatives. They share a great number of properties. They have the same shape, the same weight, and the same color. Not only do they share many intrinsic properties, but also many of the same extrinsic properties as well. For instance, if one is located in Boston, the other is. Yet they differ in other properties. The lump has the property *would survive a crushing*, while the statue does not. The lump has the property *being a lump*, while the statue has the property *being a statue*. Some people have also argued that the statue has certain aesthetic properties that the

---

[20] Building a model is not the same thing as making a metaphysical commitment. Still, the various metaphysical responses to the statue and the lump undermine a default assumption many people seem to have: that all we need to do in tracking the statue is track the lump.

[21] Baker 1999; Doepke 1982, 1996; Fine 2003; Forbes 1987; Johnston 1992; Koslicki 2004; Lowe 1995; Oderberg 1996; Paul 2010; Shoemaker 1999, 2003; Simons 1987; Thomson 1998; Wiggins 2001; Yablo 1987.

lump does not. How are these two objects so closely related without being the same?

The prevailing proposal is that there is a relation—the *constitution* relation—which holds between certain pairs of objects that spatially coincide and yet are not identical. The lump constitutes the statue: the lump is the material out of which the statue is made. Constitution is widely understood to be *asymmetric:* while the lump constitutes the statue, the statue does not constitute the lump. And *irreflexive*: the lump does not constitute the lump, nor does the statue constitute the statue. Thus constitution is different from mere spatial coincidence, which is symmetric and reflexive. (If *a* coincides with *b*, then *b* coincides with *a*. And everything coincides with itself.)

As with the grounding relation, constitution seems to involve a kind of hierarchy of "fundamentality." And as with the grounding relation, work on the constitution relation is still in its infancy. Even its proponents disagree on its characteristics. The burgeoning literature on it pursues two different questions: (1) *Can there be objects that spatially coincide without being identical?*[22] and (2) *What is the nature of the constitution relation?*[23] These are often addressed together, but are really separate issues.

A Popular Account of Constitution

We can draw an analogy between ordinary objects, like the statue, and groups like the MassDOT board or the Supreme Court. The lump is not identical to the statue, but is the material of which the statue is made. Similarly, Alito, Breyer, Ginsburg, et al., are not identical to the Supreme Court, but they are the people out of whom the Supreme Court is made. In 2014, those particular people are the members of the Supreme Court. But the court can change its membership, so that over time it continues to exist, even without having those people as its members.

In a recent paper, Gabriel Uzquiano has given one proposal for how to apply constitution to groups.[24] Although his discussion highlights some key characteristics of constitution as applied to groups, the account itself is unsuccessful. I will propose a different approach to constitution and its application to groups.

---

[22] See Bennett 2009; Hawthorne 2006; Hirsch 2002; Paul 2002; Sider 2009.
[23] See Baker 2000, 33, 45–6; Hawley 2006; Lowe 1989, 81; van Inwagen 1987; Wasserman 2004, 694.
[24] Hindriks 2008 also applies Lynne Rudder Baker's account of constitution to a kind of social group—the kind he calls "corporate agents." I discuss Hindriks's view in chapter 18.

Uzquiano builds his account of group constitution on a popular account of constitution given by Judith Jarvis Thomson in "The Statue and The Clay."[25] Thomson begins by noticing that the lump—a particular portion of clay—is a different portion of clay if you take away its parts. Suppose the statue/lump is on a pedestal. If you break off pieces of the lump and scatter them on both the pedestal and the floor, then that portion of clay is scattered. After the scattering, the portion of clay left on the pedestal is a different portion. On the other hand, if you break a couple of fingers off the statue and drop them on the floor, then the statue on the pedestal is still the same statue. In other words, not all the parts of the statue are essential to it: if you take a piece away from the statue, or add to it, it can remain the same statue. But if you take a piece away from the lump, or add to it, it is no longer that lump.

Thomson uses these characteristics to develop an analysis of what it is for the lump to constitute the statue at time t. The first condition captures the fact that they coincide at t:

(10.2) The statue and the lump have the same parts at time t.

Then, to capture the idea that the statue can change its parts while remaining a statue, whereas the lump cannot change its parts while remaining a lump, she adds two rather complicated conditions:[26]

(10.3) There is some part x that the lump necessarily has, while nonetheless x is not necessarily a part of the statue, nor is any part of x necessarily part of the statue. (Putting it roughly: some parts that are essential to the lump are not essential to the statue.)

(10.4) There is no part y that the statue necessarily has, but for which y itself does not have at least some part that the lump necessarily has. (Roughly: All the essential parts of the statue have parts that are essential to the lump.)

The point of these conditions is that the lump is more tightly connected to its parts than is the statue. In Thomson's conception, the constitution relation holds between two objects when they have the same parts at a given time, but the constituted object can change its parts more readily than the constituting object can. With these conditions, Thomson characterizes a relation that is more than just spatial coincidence. Coincidence is reflexive and

---

[25] Thomson 1998.
[26] Thomson 1998, 157.

symmetric. Thomson's constitution relation, on the other hand, is asymmetric and irreflexive.

Applying Constitution to Groups

Like Thomson, Uzquiano characterizes the relation between a group and its membership by contrasting the tightness of each entity's relation to its parts. Only, Uzquiano has to change the account a bit, because with groups, it is not quite right to speak about parts, and parts of parts. One of the parts of Samuel Alito, for instance, is his right arm. But it does not seem right to say that that Samuel Alito's right arm is one of the parts of the Supreme Court. (Or at least, if Alito's right arm does count as a part of the Supreme Court, then "parthood" is not the relation we are interested in. Rather, the relevant relation is that Alito is a *member* of the Supreme Court—and Alito's arm is certainly not that.) Thus when we characterize the coincidence of Alito, Breyer, et al., and the Supreme Court, Thomson's condition (10.2) is not the best way to go. Instead, we need to identify what, for the case of groups, the analogous object is to the lump. Uzquiano's solution is to take a group like the Supreme Court to be "group-constituted" by a *set*, such as the set {Alito, Breyer, Ginsburg, Kagan, Kennedy, Roberts, Scalia, Sotomayor, Thomas}.

Sets Again?

Earlier in this chapter, I pointed out that it is a mistake to *identify* a group like the Supreme Court with a set of people like {Alito, Breyer, Ginsburg, Kagan, Kennedy, Roberts, Scalia, Sotomayor, Thomas}. But Uzquiano is not doing this. He is using sets in a different way. Even though groups and sets are distinct objects, at a given time, he takes the "material" or "stuff" of a group to be the same as that of the set. A set is a thing that has its elements essentially, just as a lump has its parts essentially. Sets are not groups, but on his view sets "group-constitute" groups.

To clarify this: we can distinguish two questions about the relation of sets to groups—

1. What *are* groups? Are they simple sets, complex sets, graphs, aggregates, collections, or something else? This is the question I discussed above.
2. Presuming that groups are not sets, what entities *constitute* groups? Are groups constituted by sets, or are they constituted by graphs, aggregates, collections, or something else?

Uzquiano takes sets to be the answer to the second question. Groups are not identical to sets, but they *are* constituted by sets. To be honest, I am not confident that this is the best decision. Just as it is odd to *identify* a group with a mathematical object, it is similarly odd to have an object like a group *constituted by* a mathematical object. On the other hand, sets are precise, at least. Furthermore, whatever sorts of objects do constitute groups, they will be very similar to sets, since sets resemble lumps in many ways.[27] So, without putting too much stock in it, I will follow Uzquiano in this choice.

Uzquiano's Analysis of "Group-Constitution"

Uzquiano's task is simpler than Thomson's, since he does not need to worry about parts of parts, like Alito's arm. So he is able to propose a more straightforward analysis of "group-constitution." He gives two conditions for a set S to "group-constitute" the Supreme Court at time t:[28]

(10.5) A person x is a member of the Supreme Court at time t if and only if x is an element of S.

(10.6) There is a person x such that: (a) x is a member of the Supreme Court at t, (b) x is necessarily a member of set S, and (c) x is possibly not a member of the Supreme Court, at some other time t' when the Supreme Court nonetheless exists.

As with Thomson's conditions, the first is meant to capture the coincidence of the Supreme Court and set S at time t. Instead of talking about people as parts of groups and sets, he correlates being a member of the group with being an element of the set. Uzquiano's second condition captures the difference between how the group is tied to its members, and the set to its elements. Whereas the elements of S are fixed, the members of the Supreme Court can fluctuate over time and possibilities.

Uzquiano's aim is to analyze what more it takes for S to constitute the Supreme Court, beyond coinciding with it. Unfortunately, a close look shows his analysis to be flawed. Once we take into account what Uzquiano has already stipulated about sets and the Supreme Court, we can see that Uzquiano's condition (10.6) does not actually contribute anything at all to his definition of "group-constitution."

---

[27] In particular, in having their elements essentially.

[28] Uzquiano 2004, 151–2, gives a modified statement of group-constitution for the general case, but the differences are not important for our purposes. Also, his modification is flawed: it rules out groups whose members *are* essential to them.

Notice that Uzquiano's "group-constitution" relation is less general than Thomson's proposal. Thomson takes constitution to be a relation between an object and another object. Uzquiano takes group-constitution to be a relation between a set and a group. That is: the first relatum is always a set, and the second relatum is always a group. The problem is that we have already taken it for granted that the elements of sets cannot fluctuate, while the members of groups can. So if the first relatum is a set and the second relatum is a group, then we can logically derive (10.6) from (10.5). (See the note below for a proof of this.[29]) Satisfying (10.5) is all it takes to guarantee that S "group-constitutes" the Supreme Court at t, according to Uzquiano's definition. It is true that "group-constitution" is asymmetric, but the asymmetry is already present in (10.5): sets do not have members, nor do groups have elements. Since Uzquiano's condition (10.6) does not add anything, his definition does not actually distinguish *s constitutes g* from *s coincides with g*.

A "Grounding" Analysis of Constitution

This does not mean it is a bad idea to apply constitution to groups. But it does mean we need a better analysis. Fortunately, another look at groups points the way to one, and to a better analysis of constitution altogether.

The flaw is not just with Uzquiano's view. Thompson's treatment of constitution does not really confront the basic insight and burden of the question. How can two ordinary objects, or two groups, coincide and yet be distinct? To explain this, we have to consider not just *that* they may relate differently to their respective parts or members, but *why* they relate in the way they do. Not all coinciding entities stand in a constitution relation to one another, but this is *not* because one is more tightly coupled to its members than another is. We can see this with the case of groups.

Consider, for instance, the three different coinciding entities (1) the MassDOT board, (2) the MBTA board, and (3) the set of people M who are their actual members. These do not all relate to one another in the same way. The MassDOT board is constituted by M, its set of actual members, and so is

---

[29] Here is a proof. Three principles are in the background: (1) sets cannot change their members, that is, for all sets S and all objects x, if x is an element of S, then necessarily x is an element of S; (2) the Supreme Court can change its membership: For every object x there is possibly a time t' such that the Supreme Court exists at t' and x is not a member of the Supreme Court at t'; (3) the Supreme Court has some members at t. Therefore, by (3), there is a person x who is a member of the Supreme Court at t, who by (10.5) is also a member of S, and who by (1) is necessarily a member of S, and yet who by (2) might not be a member of the Supreme Court even at some time it exists. This is just (10.6).

the MBTA board. But while the MassDOT board coincides with the MBTA board, they do not constitute one another. (For one thing, constitution is asymmetric. So if the first constitutes the second, the second cannot constitute the first. Moreover, neither the MassDOT board nor the MBTA board is more "fundamental" than the other.) There is a reason they have the same members, but the reason is not that one is built out of the other.

Doepke's Proposal about Constitution

In a 1996 book, Frederick Doepke proposes an account of constitution based on a different idea than the Thomson-style approach. For an object a to constitute an object b, according to Doepke, certain properties of a need to *explain* both the existence of b and b's persistence conditions.

When a constitutes b at time t, according to Doepke, a has some property P, which is accidental (i.e., one that a has contingently, not necessarily). And that property does two things. First, the fact that a has P at t explains the fact that b exists at t. Second, if a continues to have P, that explains b's persistence after t. On the other hand, facts about b are irrelevant to a's persistence. It is this *explanatory* relation, according to Doepke, that distinguishes constitution from mere coincidence. When a constitutes b, facts about a serve to explain key metaphysical facts about b. Mere coincidence does not do this.[30]

In a 2004 paper, Ryan Wasserman points out fatal flaws with Doepke's analysis. He takes Doepke to task for making P do all that explaining. For the lump to constitute the statue, on Doepke's account, some property of the lump needs to explain both the existence and persistence conditions of the statue. It is not clear, Wasserman points out, what property of the lump could do this. This is just too much to ask of a modest lump of clay.[31] Wasserman concludes that Doepke's approach is a nonstarter.

But although Wasserman's critique is correct, his conclusion is not. Doepke is not wrong to see metaphysical explanation as the heart of the difference between constitution and mere coincidence. It is simply that Doepke is too enthusiastic about what gets explained. Wasserman is right that facts about a cannot explain the identity and persistence of b. But Doepke should not have insisted on those. We can easily see this by using groups as an example.

---

[30] Doepke 1996, 200–201.
[31] Wasserman 2004, 698–9. Wasserman also points out problems if we allow P to be just any extrinsic property.

A New Proposal

First, let me replace Doepke's talk about "explaining" with something more specific: the grounding relation. What is distinctive about *a constitutes b* is that certain facts about *a* ground a certain fact about *b*.

Consider again the example of the relation between set M and the MassDOT board. It is a mistake to think that the fact **The MassDOT board exists** should be grounded—even partially—by facts about M. That board exists for independent reasons, whatever the facts about M in particular. Nor do facts about M ground the *persistence* of the board at some time. Nor do they *anchor the conditions* under which the MassDOT board exists or persists. Whatever does account for the existence and persistence of the board, the actual members of M have little if anything to do with it.

Some facts about the board, however, do rely on M. Facts about M are relevant at least to partially grounding certain facts about the MassDOT board. The MassDOT board, for instance, has certain properties, like its weight and shape,[32] which is partially grounded by the fact that M has those same properties.[33] (As we will see, even these are not *fully grounded* by facts about M. This turns out to be an important point.)

Certain facts about M partially ground certain facts about the MassDOT board, and not the other way around. The reason, one might say, is that M constitutes the MassDOT board, while the MassDOT board does not constitute M. But we do not need to leave things at that. Instead, we can explicate the constitution relation by focusing on one particular fact. Consider the following fact about the MassDOT board:

(10.7) *At t, the MassDOT board has the membership it does*.

This might look trivial, but it is not. It is not the fact **At t, the members of the MassDOT board are the members of the MassDOT board**. Rather, it is the fact **At t, the MassDOT board coincides with a particular set**, that is, the particular set that it happens to actually coincide with at t.

What grounds (10.7)? Many things. The membership of the MassDOT board involves facts about nominations, elections, appointments, resignations, and more. Still, *among* the grounds of (10.7) are facts about M. Part of

---

[32] Facts about M also partially ground certain extrinsic properties holding of the MassDOT board as well. For instance, the fact *The MassDOT board weighs more than me* involves an extrinsic, rather than intrinsic property. See Koslicki 2004 for many examples along these lines.

[33] Here is a clear problem with sets as the things that constitute groups, because sets are plausibly abstract objects. It is perhaps preferable to replace sets of people with collections of people, but again, an adequate treatment would take us too far afield.

what grounds (10.7), for instance, is the existence of each individual person in M. That is not all that grounds (10.7), but it is a part.

Notice that the reverse does not hold. The fact that M has the elements it does is not grounded—even partially—by facts about the MassDOT board.

The moral is this: constitution is not about one thing being ontologically related in the right way to another thing's existence, or to another thing's persistence. Rather, it is about the stuff of one thing being part of the metaphysical reason that another thing is made up of the stuff it is. The set of people that constitutes a group can do *some* explaining of facts about the group itself. But there are many facts about the group it does not explain.

For M to constitute the MassDOT board at *t* is for M to meet two conditions: M needs to coincide with the MassDOT board at *t*, and certain[34] facts about M need to partially ground the fact that the board has the membership it does, at *t*. This requires a lot less than Doepke suggests. But, like Doepke's proposal, it takes metaphysical explanation to be at the heart of constitution, and it takes constitution to be more than mere coincidence.

A Broad Conception of Groups

I began this chapter by stating how I will understand groups: *a group is a thing constituted by and only by individual people*. This is not a trivial characterization. In analyzing groups with the constitution relation, I am highlighting three things. Groups can coincide—distinct groups can have the same members. Groups, like ordinary objects, can persist through change. And a basic fact that needs to be grounded about a group is the fact about what set of people the group coincides with at a time.

All this contrasts with the prevailing paradigms of groups in the social science literature. It is easy to be tempted into an excessively narrow picture of how to distinguish groups from one another, and how facts about them are grounded. Using the broad conception of groups I have put forward, we can carefully discriminate various kinds of facts about groups. If we treat every group that has the same members as the same group, then we cannot even get a start on understanding why a group has the basic properties it does.

I will return in later chapters to the shortcomings of narrow conceptions of groups, in particular of the fourth paradigm: the one that currently prevails in philosophies of group action and intention. But in the next chapter I turn to a more immediate concern: what are some basic facts about groups, and how are they grounded? How do we work out the frame principles for facts about even one particular group?

---

[34] There need to be restrictions on the "facts about M." The fact **M constitutes the MassDOT board**, for instance, cannot be included. This needs further conceptual work and clarification.

11

Simple Facts about Groups

If you want to know something about a group, look at its members. Facts about a group are exhaustively determined by facts about the people who constitute it. To many philosophers and social scientists, these claims have seemed obvious. Only, they are not true.

Remember Virchow, from chapter 3. Virchow's cell theory was on target for a small number of anatomical facts that are fully grounded by facts about cells. But for other anatomical facts he was entirely wrong: their grounds do not involve facts about cells at all. And for most anatomical facts, he was partly right and partly wrong: they have hybrid, heterogeneous grounds. Most anatomical facts are partially grounded by cellular facts and partially by noncellular ones.

The same holds for facts about groups. Perhaps a few are grounded exclusively by facts about individual people. But many are grounded by facts that do not involve individuals at all. And most facts about groups are hybrids: they have heterogeneous grounds, some about individuals, and some not. This may seem surprising. After all, it is natural to think of a group as nothing but its members. But just because a group is *constituted* exclusively by people does not mean that facts about those people (or about any people) ground most facts about the group.

How are facts about groups grounded? This is the question I begin to address in this chapter. Let's consider some facts about the Supreme Court.

Simple Facts about the Supreme Court

Imagine we are living in the late 1780s, and have been appointed as delegates to the Constitutional Convention, or as members of the newly formed Congress. And suppose we need to think through the practical details of the judicial system of the United States. Some of these we will eventually write into the US Constitution, and some into the first Judiciary Act of 1789. We start with a blank slate. There is no Federal Judiciary, so we have to decide on the most basic characteristics it will have. Of course we have other judiciary systems as models, so we can weigh the pros and cons of alternative systems in designing our own, but still, it is our job to set up our own system.

We decide to establish a hierarchy of courts, with one Supreme Court at the top. But the *existence* of the Supreme Court is only the first step: we also need to establish the grounding conditions for other facts about it. For instance, we decide on the conditions for membership. Some of these conditions are specific to the Supreme Court, such as the lifetime term of justices. And some are membership conditions shared by other federal offices, such as the process of nomination by the president, and also the requirement that members be bound by an oath of office. We also realize that it is impractical for the court to be in session throughout the year. So we write into the Judiciary Act the provision that there are two sessions. Maybe we decide that one session starts in February and one in August, or maybe we decide that the court's activation should be partly grounded by the Chief Justice's choice.

We also need facts about what the court has the power to do. So we write into the act that it has appellate jurisdiction over other courts, that it has exclusive jurisdiction over matters between states, over proceedings against public ministers, and so on. We debate whether or not there should be a hierarchy among the members of the court. We decide on a chief justice and five associate justices, with precedence among them according to their dates of commission. And so on. Our decisions, of course, are not quite enough to anchor the grounding conditions for these kinds of facts. The process of enacting the laws, or ratifying the constitution, does the anchoring. My point here is to note that there are lots of different frame principles that need to be anchored.

With all these different facts about the Supreme Court in place, let's take a systematic look at some grounding conditions for particular facts. Table 11a lists some examples of facts about the Supreme Court, sorted into rough categories.

Table 11a **Some facts about the Supreme Court**

Constitution	(SC1) *The Supreme Court is constituted by the set {Alito, Breyer, Ginsburg, Kagan, Kennedy, Roberts, Scalia, Sotomayor, Thomas}*
Existence	(SC2) *The Supreme Court exists.*
Hierarchy	(SC3) *John Roberts is chief justice.* (SC4) *The associate justices are arranged in such-and-such a hierarchy of precedence.* (SC5) *When the chief justice is unable to perform the duties of office, or the office is vacant, the powers and duties devolve on the associate justice next in precedence.*

Table 11a *Continued*

Activation	(SC6) A quorum is present.
	(SC7) The Supreme Court is in session.
	(SC8) All courts of the United States are always open for the purpose of filing proper papers, etc.
Actions	(SC9) The Supreme Court voted in favor of Citizens United.
	(SC10) The Supreme Court entered the chamber at 10 a.m. this morning.
Preferences	(SC11) The Supreme Court wants to keep video cameras out of the courtroom.
Intentions	(SC12) The Supreme Court intends to issue fewer writs of certiorari than it did in past decades.
Physical facts	(SC13) The Supreme Court is now in Washington, DC.
	(SC14) The Supreme Court is sweating in the August humidity.
Jurisdiction	(SC15) The Supreme Court has exclusive jurisdiction over such-and-such cases.
	(SC16) The Supreme Court has original but not exclusive jurisdiction over such-and-such cases.
	(SC17) The Supreme Court has appellate jurisdiction over such-and-such-cases.
Powers	(SC18) Cases in the courts of appeals may be reviewed by the Supreme Court in such-and-such cases.
	(SC19) The chief justice has such-and-such powers.
	(SC20) Associate justices have such-and-such powers according to the hierarchy.
	(SC21) In such-and-such circumstances, the Supreme Court has the power to strike down laws as unconstitutional.
	(SC22) The Supreme Court has the power to issue such-and-such writs.
	(SC23) The Supreme Court has the power to grant new trials under such-and-such circumstances.
Obligations	(SC24) In such-and-such circumstances, the Supreme Court shall not issue execution in such-and-such cases, but shall issue a mandate to the circuit court to award execution.
	etc.

This is a long list, and still does not include any economic facts, historical facts, ethnographic facts, and so on about the Supreme Court. It does, however, give us an idea of the range of facts whose grounds we need to work out.

Let me underline this point. You will often find, when examining a theory in social ontology, that it addresses only one or two kinds of facts. Maybe you will find a theory of facts about group action, or a theory of facts about group membership. But, for the most part, that is it. Notice, however, all of the facts we just listed about the Supreme Court. From this list alone we can see that the facts about even one simple group can be dizzyingly diverse. If we want our theory of groups to have practical applications, we cannot shrink away from this rich terrain. Moreover, once we expose ourselves to all the different kinds of facts about social groups, we can see new patterns emerge—patterns we would be blind to, were we to limit our vision.

Rooting Out the Grounding Conditions for Specific Facts

Normally, it takes a lot of work to think through the grounding conditions for facts about a group. But here, the Supreme Court example has some advantages. It is an unusually conspicuous group, important enough that the grounding conditions for many facts about it are codified explicitly in the law books.[1] Generations of lawyers and legislators have done our work for us, sorting out grounding conditions. So we can exploit their labor.

I will begin with the first fact on the list:

(SC1) *The Supreme Court is constituted by the set {Alito, Breyer, Ginsburg, Kagan, Kennedy, Roberts, Scalia, Sotomayor, Thomas}*

What other facts fully ground this one? We should start with the conditions for individual membership in the Supreme Court. These are explicitly recorded in the US Constitution and various judiciary acts. Roughly speaking, a person x becomes a Supreme Court justice just in case the following sequence of conditions is satisfied in the following order:

---

[1] I should caution that legal facts are not the central or motivating examples for the treatment of social facts in this book. Legal facts are useful because their frame principles are so explicit. But this can be misleading. It is atypical for a group to have so many of its frame principles so explicitly anchored.

1. There is a vacancy on the court
2. The president nominates *x* as a candidate
3. The Senate confirms *x*
4. *x* takes the oath of office

In thinking through the conditions for being a Supreme Court justice, it is natural just to think about the conditions for *becoming* a justice. But, of course, these are not quite sufficient—it is possible to stop being a justice as well. For membership in groups, it is often helpful to divide the conditions into three parts:

> *Initiation conditions*: the conditions a person must satisfy to become a member
> *Maintenance conditions*: the conditions for remaining a member, once initiated
> *Exit conditions*: the conditions for being removed as a member.[2]

Some groups have significant maintenance conditions. To be a member of a health club, for instance, one might have to pay an initiation fee and then pay ongoing monthly membership fees. Remaining a Supreme Court justice is easier than remaining a member of a health club. Even if Samuel Alito fails to come to court for a whole term, he remains a Supreme Court justice. As far as I can tell, the only maintenance condition for a Supreme Court justice is that the candidate be alive. There are, however, more substantial exit conditions. If a justice resigns, she is no longer a member of the court. There is also a formal process for removal of a justice, involving impeachment by the House and removal by the Senate.

Notice that the facts involved in initiation, maintenance, and exit are largely facts about people and things other than the justices. They have a bit to do with Alito, Breyer, and the others, but also depend on many other facts about the president and the Senate as well. To see this, suppose former Justice Sandra Day O'Connor is so upset at recent Supreme Court decisions that she has a mental breakdown. She forgets that she resigned back in 2006, and comes to believe that she is still a justice. Suppose all her beliefs return to what they were back in 1985—she honestly thinks she is a justice, dresses for work, and shows up at the Supreme Court door. Regardless of her beliefs, she is not a Supreme Court justice. What makes someone a member of the court is that the person satisfies all kinds of external conditions, which O'Connor does not. The same is true even if

---

[2] It is helpful to think of the maintenance and exit conditions separately: the maintenance conditions are those one needs to satisfy on an ongoing basis to stay a Justice, and the exit conditions are those that take away the position, even if the maintenance conditions are satisfied.

every member of the Supreme Court has a mental breakdown. Suppose all of their memories are reverted to what they were in 2006. O'Connor might then show up at the Supreme Court door, and be let in, and sit at the bench, and be greeted by the other people at the bench, and so on. Still, she would not be a Supreme Court justice, despite the insistence of everyone at the bench.

As I said above, other facts are involved in grounding facts about membership, apart from those about the members alone. Some of these facts can be easy to overlook. The first "initiation" condition, for instance, is that there must be a vacancy on the Supreme Court. This is a fact about other people altogether, apart from the current membership of the Supreme Court, the president, and the Senate. There is a vacancy only if *other* people in the population do not already satisfy the initiation, maintenance, and nonexit conditions. Another grounding fact that is easy to miss is that the Supreme Court exists in the first place. As I discussed in chapter 8, a fact of the form ***a* has property P** is not just a fact about P, but also about the object *a*. With no Supreme Court, a person might satisfy all the prior conditions, and yet fail to be a Supreme Court justice.

We might depict all the grounding facts, then, with figure 11A:

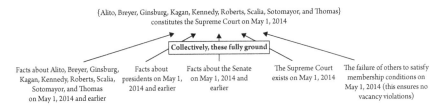

Figure 11A Grounds of the constitution of the Supreme Court

Some of these grounding facts may seem finicky. But it pays to be finicky, when we are talking about the most basic and foundational facts of all. Nearly any other fact about the Supreme Court we might be interested in depends on facts of the form *y constitutes the Supreme Court*. Nearly any fact about the Supreme Court, that is, depends on *every single one* of the grounds of the types listed in figure 11A. How did the Supreme Court vote over time? What was the influence of political pressure or lobbying groups on the Supreme Court? What are the Supreme Court's spending patterns and educational attainments? To answer questions like these, we investigate facts that depend on facts about the Supreme Court's constitution. This means that all the grounds listed in figure 11A, among many others, may be pertinent to practical facts we are interested in. Changes in *any* of these various grounding facts thus have the power to change nearly any fact we might be interested in about the Supreme Court.

To write a frame principle for facts like (SC1) more explicitly, we need to wrap all this up in a single set of grounding conditions. To abbreviate, I will use Q to symbolize the property of satisfying the initiation, maintenance, and removal conditions for membership in the Supreme Court. Most of the grounds listed in figure 11A are included in this property Q. For instance, I built into the initiation conditions for each member the "vacancy" condition that there be an empty slot for that person to fill. The only ground missing from Q is the fact **The Supreme Court exists at t**, so that fact should be explicitly mentioned as a grounding condition in the frame principle.[3] We also need to write the whole thing as a condition that a *set* of people satisfies as a whole, not just the various members. (This is because the constitution relation is usually understood as a relation between two objects—one constituting entity and one constituted entity.) So we need to combine the membership conditions for individual justices into a single condition that a set of people y must have if it is to constitute the Supreme Court.

Altogether, we can write a frame principle for determining facts about the Supreme Court's constitution as:

(11.1)   It is necessary* that: for all y, and for all times t: if the Supreme Court exists at t and y is the largest set of people each of whom has Q at t, then those facts ground the fact that y constitutes the Supreme Court at t.[4]

I have marked 'necessary' with an asterisk, to remind ourselves that what follows it is a statement of the grounding conditions for the social fact in every world *in the frame*.[5]

At heart, frame principle (11.1) is a conditional: if fact G obtains, then fact F obtains. Or, to be more precise, it is a determination relation: necessarily, if G obtains, then G grounds the fact that F obtains. In this case, the "grounding condition" is **the Supreme Court exists at t and y is the largest set of people**

---

[3] It may be that this fact actually is included, as part of the grounds for there to be a vacancy. But it is useful to keep existence explicit as a ground as well.

[4] The condition "the largest" is included so as to ensure that all of the justices are in the constituting set.

[5] Three points about this. First, this expression should be interpreted as evaluated at a frame: at every world in the frame, if G then G grounds F. In an elaborated model of multiple frames, we could distinguish frame principles at different frames by using subscripts or some other marker to indicate the frame to which the principle applies.

each of whom has Q at t. The "grounded fact" is **y constitutes the Supreme Court at t**. Thus (11.1) gives a set of contingent grounding conditions for social facts of a certain type.

Many frame principles fit into this sort of template: they give one type of contingent fact that determines a type of social fact. But this is not the only form frame principles come in. Some frame principles, for instance, have *necessary* facts as grounding conditions for a social fact. Or, as we will see, a frame principle can even have no grounding conditions at all—that is, where it is simply anchored that some fact obtains in every world in the frame. While (11.1) is a typical example, frame principles come in many forms.

Frame principle (11.1) is about the facts that *determine* the constitution of the Supreme Court. It is not a principle about dependence. It does not capture the fact that this is the *only* way to ground the constitution of the Supreme Court. Some social facts might have a variety of determination principles. (Think, for instance, of the ways a fact like **Bob and Jane are married** could possibly be grounded: signing a paper at city hall, going through a religious ceremony, etc.) In the present case, however, (11.1) gives the only way for a set to constitute the Supreme Court. So it is easy to add a dependence principle as well:

(11.2) It is necessary* that: for all y, and for all times t: if y constitutes the Supreme Court at t, then that fact is grounded by the facts that the Supreme Court exists and that y is the largest set of people each of whom has Q at t.

These frame principles apply to all the possibilities in the frame. They give us a template for investigating various facts that are eligible to change the grounded fact. If any one of the conditions in the frame principle fails to be satisfied by a set y in a world at a time t, then in that world, the fact **y constitutes the Supreme Court at t** does not obtain.

Above, I said that I was giving a frame principle for a particular fact. But really, (11.1) and (11.2) give the grounding conditions for a whole range of facts. That is, they do not just give us the grounding conditions for the one fact **{Alito, Breyer, . . . , Thomas} constitutes the Supreme Court in 2013**.

> Second, tied to what I have said about "contingency" vs. "necessitarian" views of grounding, there might be certain frame principles that do not apply to all the worlds in the frame. This is remote from present considerations, so for our purposes we will take all frame principles to involve necessity*, that is, to hold for all worlds in the frame.
>
> Third, the conjunctivist can just regard the frame principles to be restricted to the worlds in which the anchors obtain.

Instead, they are general frame principles for *any* set *y* that might constitute the Supreme Court, at any time, and in any possibility in the frame.

Altogether, it is not trivial to work out frame principles for a single fact about a group, but it is not particularly mysterious or difficult either. With this example, we see some basic forms frame principles can take, and also that the grounding conditions involve much more than just facts about the members.

Existence

Consider the conditions for an even simpler fact: **The Supreme Court exists at *t*.** Interestingly, these grounding conditions are not explicitly laid out in the Constitution or anywhere else. It is not even obvious how or when the *actual* Supreme Court came to exist—to say nothing of the Supreme Court's existence conditions in other possibilities. Article III of the Constitution says that the "judicial power of the United States, shall be vested in one supreme Court," but leaves it to Congress to work out the details. The Judiciary Act of 1789 says "the Supreme Court of the United States shall consist of a chief justice and five associate justices, etc." The executive is implicitly instructed to appoint justices.

Did the Supreme Court exist, with no members, when the Constitution was ratified, or when the Judiciary Act of 1789 was passed? Does it *necessarily* exist in the frame—that is, did the framers of the Constitution anchor a frame principle so that the Supreme Court's existence is fixed throughout the frame? (As I mentioned, frame principles do not have to have contingent facts as grounding conditions. Or even any grounding conditions at all. Sometimes anchoring just puts in place facts about social objects in every world in the frame.) Or rather, is the Supreme Court's existence grounded by some contingent facts? Did it come to exist with the inauguration of its first member? Or under some other conditions?

On their own, these questions may seem rarefied or merely academic. However, they raise three crucial points. First, the grounding conditions for the Supreme Court's existence are very different from the grounding conditions for its having the constitution it does. The frame principles for **The Supreme Court exists at *t*** are not (11.1) or (11.2). Second, frame principles in general can take a variety of forms: they need not be necessitated conditionals like those above. And the third point relates to *kinds* of groups, like courts, classes, legislatures, or markets. For groups of a given kind, it is extremely practical to consider the frame principles for facts about existence. Suppose, for instance, you were building a model of declining union activity in the last 50 years. It would be natural to include the creation and elimination of unions as a central part of the model. Or consider a model of incentives for new business formation, or for forming new and smaller classes in charter schools. The target of such a model is precisely the existence or non-existence of certain groups and other social entities. So even if the grounding conditions for *The Supreme Court exists* seem obscure, looking into them sets us up to understand an important category of facts.

How exactly to cash out the frame principle for the Supreme Court's existence will depend on whether (1) the existence of the Supreme Court is contingent on the occurrence of some event, such as the appointment of the first justice or justices, or (2) its existence is necessary in the frame. We can use this case to illustrate both alternatives.

Alternative 1: The Existence of the Supreme Court is Contingent in the Frame

Suppose the existence of the Supreme Court is grounded by the occurrence of some event. In the actual case, the first five justices were confirmed by the Senate on September 25 and 26 of 1789, and then received their commissions over the next few days. It is not obvious what event triggered the existence of the Supreme Court. Nor is it obvious whether it existed before the event without any members, or whether the first member's commission created it, or the Chief Justice's commission created it. Let us suppose that it was the first member's commission that grounded the existence of the Supreme Court.

Also relevant to its existence is whether the Supreme Court can ever go out of existence, after it has been created. Here too, it is not obvious what the answer is. For simplicity, I will assume that once it has come to exist, it continues to exist in perpetuity. In the actual case, this means that the existence of the Supreme Court at all times after its establishment is grounded by John Rutledge's commission as the first member on September 25, 1789. Its existence in 1789 is grounded by that fact, as is its existence in 1950, 2015, and the year 3000. The occurrence of Rutledge's commission grounds all these facts, but they also could have been grounded in other ways. If John Jay had received his commission first, that would have grounded the Supreme Court's existence then and subsequently.

Granting these assumptions, the fact *The Supreme Court exists at time t* is grounded by a fact of the form *At t or earlier, someone has received commission as the first Supreme Court justice.*[6] Then the determination and dependence frame principles, respectively, are:

(11.3) It is necessary* that: if someone has received commission as the first Supreme Court justice at *t* or earlier, that grounds the fact that the Supreme Court exists at *t*.

(11.4) It is necessary* that: if the Supreme Court exists at *t*, then the fact that someone has received commission as the first Supreme Court justice at *t* or earlier grounds the fact that the Supreme Court exists at *t*.

Alternative 2: The Existence of the Supreme Court is Necessary in the Frame

Suppose, instead, that the existence of the Supreme Court is necessary in the frame. Here, the relevant frame principle has a simpler form:

(11.5) It is necessary* that: the Supreme Court exists.

This frame principle is also very different from (11.1) and (11.2). On this alternative also, the grounding conditions for existence diverge from the grounding conditions for constitution: **The Supreme Court exists** has no grounding conditions. (Or, if you are a conjunctivist, as I discussed in chapter 9, then the anchors are the only grounds.) This frame principle does not even have the same form as (11.1) or (11.2). It just expresses a necessary fact that the anchors have put in place in our frame.

It may seem strange to regard the Supreme Court's existence as being necessary in our frame. I think this is not such a problem: given that we have anchored the frame as we have, it is not clear that this group's existence needs to be contingent. Still, if that seems unpalatable, there are alternatives: you could go back to alternative (1), and take the frame principles to be (11.3) and (11.4), or else go with conjunctivism and collapse the anchors into the grounds. Altogether, it is not obvious how to understand Article III's "judicial power of the United States, shall be vested in one supreme Court." Is it a description of what will be the case when certain conditions are satisfied? Is it a performative statement, an act that goes into force and anchors the

---

[6] This may seem circular, since receiving that commission might require that the Supreme Court exists. But both of these facts may be grounded in the same process. Or it may be that the commission is for the as-yet-to-be-created Supreme Court.

existence of the court when the Constitution is ratified? Interesting questions, but neither hill seems worth dying on. It is when we move to *kinds* of groups that existence conditions become really interesting. There we see that the existence of groups—like their constitution—can be grounded by all kinds of facts, far beyond facts about their members. As we will see, however, when we turn to kinds in chapter 13, frame principles for existence get a bit complicated. This is why I began with the simpler case of the Supreme Court.

A Structural Flaw in Searle's Constitutive Rules

Compare the grounding conditions for ***y constitutes the Supreme Court at t***, and for ***The Supreme Court exists at t***. These facts are determined in very different ways. The fact ***y constitutes the Supreme Court at t*** is grounded by a set y having a particular property Q at t, and by the Supreme Court existing. The fact ***The Supreme Court exists at t***, on the other hand, is not triggered by a fact of that sort. The facts determining the existence of the Supreme Court—if there are any—are not the same as those determining its constitution.

This point might seem obvious. But consider, for instance, Searle's formula for constitutive rules. Whether we take them in his form—X counts as Y in C—or in a conditional—if z has property X, then z is Y—either way we conflate the fact ***Y exists*** with the fact ***Y is constituted by some particular thing z***. On Searle's view, what are the grounds for ***This piece of paper constitutes a dollar***? The fact that this piece of paper satisfies the X conditions. And what are the grounds for ***That dollar exists***? The same fact: that is, that this piece of paper satisfies the X conditions. In other words, embedded in Searle's constitutive rule is the assumption that a social object's existence conditions are *exactly the same* as its constitution conditions. But the assumption is mistaken, and therefore so is Searle's constitutive rule formula. Noticing this is an immediate payoff of taking a serious look at frame principles.

More Frame Principles

Right from the start, we see that basic facts about groups can have heterogeneous grounds, ones that have little to do with the members. Frame principles for different facts can also take very different forms. We have already seen this for facts about constitution and existence. But the same observations hold for other facts too.

It would be tedious to write out frame principles for all the facts on the list, but let's consider just a couple. Consider, for instance, (SC4), the fact **The associate justices are arranged in such-and-such a hierarchy of precedence**. That fact depends on nothing more than the Supreme Court existing. One way of expressing a frame principle for (SC4), then, is:

> (11.6) It is necessary* that: if the Supreme Court exists, that grounds the fact that the associate justices are arranged in such-and-such a hierarchy of precedence.[7]

This formulation is nothing special: there is no single privileged way of expressing a frame principle. We could, for instance, describe the hierarchy of precedence as conditions on the individual members, rather than as an overall fact about the Supreme Court. Then we could capture the same hierarchy with principles of the form:

> (11.7) It is necessary* that: if the Supreme Court exists, and x is a member of the Supreme Court, and x was the first appointed among the current associate justices and x is not the chief justice, then those facts ground the fact that x is first in precedence.

Many facts about the Supreme Court involve powers, norms, rights, and obligations. One of the powers of the Supreme Court, for example, is (SC20), the power to strike down a law as unconstitutional.[8] This frame principle can be expressed as grounding a power, particular exercises of which are grounded by particular facts:

> (11.8) It is necessary* that: If the Supreme Court is in session at t, L is a law at t, a case has been brought before the Supreme Court involving law L in such-and-such a way, those ground the fact that at a later time t', the Supreme Court can strike down L as unconstitutional.

Or more illuminatingly, it might be cashed out in terms of the conditions for that power to be exercised:

---

[7] For clarity, I leave out the reference to time.
[8] Interestingly, this power may be partially anchored by English precedent; see Prakash and Yoo 2003.

(11.9) It is necessary* that: If the Supreme Court is in session at t, L is a law at t, a case has been brought before the Supreme Court involving law L in such-and-such a way, and at a later time t' the Supreme Court votes in favor of the proposition that L is unconstitutional, then these facts ground the fact that subsequent to t', L is not a law.[9]

My point is to illustrate how heterogeneous the grounds can be for even simple facts about the Supreme Court. I have described the actual grounding conditions for *The Supreme Court exists*, *The Supreme Court has such-and-such a membership*, *The Supreme Court is in session*, and other facts. These grounding conditions are not arbitrary: they were chosen for good reasons. But the legislators in 1789 could have anchored these grounding conditions to be more or less whatever they wanted. The grounding conditions could include the moon being in a particular phase. Or herring movements in the North Atlantic. Or the timing of the tides in Nantucket. Merely because the fact is a fact about a group does not mean that the grounding conditions are facts about its members, or about any individuals whatsoever. The grounding conditions can be all over the map.

Contrasting Natural Facts and Social Facts

Typical facts about social groups are not determined in the same way that typical facts about ant colonies are. Ant colonies, of course, have no jurisdictions, no rights, no obligations, no powers, no "being in session." But even with simpler facts, the contrast is evident. Facts about the constitution and existence of ant colonies, for instance, mostly depend on facts about the ants. Whereas facts about the constitution and existence of the Supreme Court could depend on facts about the phases of the moon, should we anchor them in that way. (That may sound ridiculous. But actually, facts about when the court is in session *are* grounded in facts about the earth's position relative to the sun.[10])

If there is any moral to the book, this may be it: facts in the social sciences are grounded differently than are those in the natural sciences. Compared to the social sciences, the ontology of natural science is a walk in the park. Even facts about constitution and facts about existence are grounded differently for

---

[9] It may also be that the power is retroactive. Then the frame principle would be: if L is a putative law at t, etc., then those facts ground the fact that L was not a law in the first place. See Bender 1962.

[10] Or, nowadays, facts about a cesium beam at the National Institute of Standards and Technology.

social groups than they are for objects in the natural sciences. This may already be clear by looking at figure 11A. To illuminate the point, however, let us consider a fact or two about a natural object.

Consider facts about some ordinary object we might treat in natural science. Something very simple: for instance, a rock. (Or, if that does not sound scientific enough, consider a quartz crystal, or a cell, or a strand of DNA.) The rock I am holding in my hand (call the rock R) is constituted by a large lump of particles $r_1 \ldots r_n$, all clustered and bonded together in various ways.[11] What grounds the fact **Rock R is constituted by lump $r_1 \ldots r_n$**? The answer to this is simpler than Supreme Court membership, but nonetheless has an interesting wrinkle.

Most of the grounds are obvious. The fact that these particles constitute the rock is grounded by the facts about them being clumped together in the right way. That is, by the facts **All the particles $r_1, \ldots, r_n$ are in the same spatial region; r_1 is bonded in such-and-such a way to r_2; r_2 is bonded in such-and-such a way to r_3**; and so on.

Those are almost enough, but not quite. Consider this. Suppose there were a thousand *additional* particles, $r_{n+1}, r_{n+2}, \ldots, r_{n+1000}$, which were also contiguous with and bonded to the others. Then the rock would include those *other* particles as well. Objects in the natural sciences are more like statues than like lumps. Their parts do not tend to be essential to them. Even so, the facts determining their constitution tend to be simpler than those for groups. For most objects in the natural sciences, spatial contiguity is *most* of what it takes to be a part of a given object.[12] For the rock to include particles $r_{n+1}, r_{n+2}, \ldots, r_{n+1000}$, all that would be needed is for them to be contiguous and bonded too. So in the original case, we need to add one more grounding fact: the fact that there are *not* such additional contiguous and bonded particles. In addition to the fact that $r_1, \ldots, r_n$, are spatially contiguous and appropriately bonded, there is also the fact **$r_{n+1}, r_{n+2}, \ldots, r_{n+1000}$ are not contiguous and/or appropriately bonded**. Nor are any other extraneous particles. In other words, this second fact is that the space surrounding the rock is appropriately empty and/or unbonded.[13] Figure 11B thus depicts the full grounds.

This may seem a bit finicky. But now consider the full grounding of a more mundane fact about the rock—a fact such as **The average temperature**

---

[11] We can regard the lump being the mereological fusion of the particles.

[12] Actually, matters can be much more complex in the natural sciences as well. Biological entities are individuated in complex ways, as are certain entities treated in physics. Still, they are not nearly as unconstrained as even the most ordinary social entity.

[13] More accurately: there is no adjacent space occupied by particles that are contiguous and appropriately bonded.

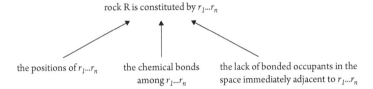

Figure 11B Grounds of the constitution of the rock

of rock R is 70°. If we were to consider what grounds this fact, the natural answer is that it is grounded by the positions, masses, and velocities of the particles $r_1,\ldots,r_n$. But again, this is not quite enough. We also need the fact that the rock is constituted by those particles. And that is not just a matter of their positions, masses, and velocities. The full grounds are depicted in figure 11C. In the figure, I have marked the intrinsic facts about the lump $r_1 \ldots r_n$ in bold type.

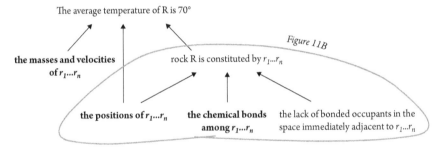

Figure 11C Grounds of the temperature of the rock

With this diagram, I want to highlight two ways this case is similar to the Supreme Court case, and one way it is not. First, part of what grounds even a simple fact like **The average temperature of R is 70°** is the fact about what constitutes the rock. As with the case of the Supreme Court, the fact of the rock's constitution is part of what grounds this fact.

Second, this means that not all of the facts that ground the rock's temperature are intrinsic. There can be changes external to the actual constitution of R that change the fact **The average temperature of R is 70°**. (I mean changes that *metaphysically ground* a change in temperature. It is obvious that external things, like a flame, can causally change the temperature of the rock. But if we change the external circumstances in such a way that we change the particles that constitute the rock, we can change the temperature of the rock without changing the position or momentum of any of the particles $r_1,\ldots,r_n$.) This too is similar to the Supreme Court—we can change facts about the court by changing facts about the president or Senate, even if the individuals remain untouched.

The third point, however, is an important contrast with the Supreme Court. *Most* of the facts that ground the rock's temperature *are* intrinsic facts about the lump $r_1 \ldots r_n$. For rock R, the facts about its material constitution are almost fully grounded by facts about $r_1, \ldots, r_n$ on their own. This means that when we want to construct a model of the properties of the rock, we can usually keep the nonintrinsic factors in the background. To put it more concisely, if more technically: the rock is nearly, though not completely, *intrinsically individuated*. To find the factors that make the rock what it is, and that make it have the parts it does, you do not need much more than the material constitution of the rock itself.

This is fairly common for objects treated in the physical sciences, like rocks and planets and cells. Typically, they are nearly intrinsically individuated. *In the social world, however, it is the rare exception.* Social objects only rarely even come close to being intrinsically individuated. There is a striking contrast between entities like rocks and entities like the Supreme Court. Just a glance at figure 11A shows that most of what determines membership in the Supreme Court are facts about things other than the justices themselves.

It usually does not do too much damage to assume that objects in the natural sciences are intrinsically individuated. The social sciences are a different story.

This Phenomenon Is Not "Emergence"

We can fix all the individualistic facts about Alito, Breyer, Ginsburg, and the others, and yet not fix simple facts about the Supreme Court. The widely assumed supervenience claims fail spectacularly, even for facts like the existence and constitution of the Supreme Court.

Some people, I have found, draw the wrong conclusion when this is pointed out. They take it to demonstrate something about "emergence." It is important to clear up this misapprehension. An example of an "emergent" property might be the sloshing of water in a cup, given rise to by the interactions of a septillion water molecules. Here the collection of water molecules has properties that none of the individual molecules have. The collection has properties that cannot easily be deduced, even if we have deep knowledge about the properties of the individual molecules.

Like glasses of water, large groups of people may have emergent properties. A group of thousands of people interacting may be like the school of herring I described in chapter 2. A moving mob may exhibit characteristics, like splitting and forming vacuoles, when it encounters obstacles. Or there may be "aggregation reversals," like the paradox of thrift: a large number of people all saving their earnings may result in less savings for everyone.

That, however, is not the point of the preceding pages. Emergence is *compatible* with supervenience. The point we have just discussed, on the other hand, is the *failure* of supervenience. For instance, the fact **The Supreme Court is constituted by {Alito, Breyer, . . . , Thomas}** is *not* exhaustively grounded by individualistic facts about the members alone. To give the full grounds of this fact, we need to look beyond the members themselves. That fact does not "emerge" from facts about the members.

Why Are the Grounding Conditions So Heterogeneous?

Human social groups are distinctive. They are not merely complicated ant colonies. We have great flexibility about which grounding conditions we anchor for even the most basic facts about groups. Most facts about social groups are *not* determined merely by facts about their members.

Why so? Because it is immensely productive for us to be able to pin the obtaining of social facts about groups to particular material circumstances. Consider, for instance, the variety of grounds that figure into a group having jurisdiction over a given activity or case. There are many groups that have enforcement powers: the marshals, the sheriffs, police forces, the FBI, the DEA, the CIA, the FDA, the USDA, and so on. The sheriff has jurisdiction within the borders of the county, and the police chief within the borders of the city. The coast guard has jurisdiction in US waters, and under certain conditions has jurisdiction in foreign waters as well. And so on. These jurisdictions are activated in different places and at different times. If the sheriff crosses over the county line, his jurisdiction lapses, even if neither he nor the perpetrator he is pursuing knows it.

Similarly, the FDA has jurisdiction over certain activities involving shelled eggs, while the USDA has jurisdiction over activities involving liquid, frozen, and dehydrated eggs. The USDA regulates sausage meat, and the FDA regulates sausage casings. The grounds for facts about jurisdiction are entirely unfettered, and similarly for facts about the activation, membership, and even existence of groups of various kinds.

Our ability to anchor social facts to have nearly arbitrary grounds is the very thing that makes the social world so flexible and powerful. Why would we deprive ourselves of that flexibility? This is what would be done by insisting that facts about groups be grounded only by facts about individuals, or still worse, by facts about the very people who are the members of those groups.

12

The Identity of Groups

Suppose you are at the Tufts basketball court, watching your roommate's intramural team play. It is late in the fourth quarter, and your friend's side is down 72–21. You remember watching them get a similar shellacking a month ago, and wonder if the team they are playing tonight is the same team as before. To decide this, you cannot just look at the players on the court. The other team may have rotated different players in tonight, or some players may have joined or dropped in the interim. Tracking a team over time is not the same as tracking a set of people.

Keeping track of a group, like an intramural basketball team, is not just an abstract exercise. We implicitly do it all the time. We track sports teams over many seasons. We track them as they persist, even as players, managers, and even team names change. Often, we need to look back in history in order to distinguish one group from another. A group can change its membership over time and in different possibilities, while remaining the same group. And distinct groups can coincide—they can have exactly the same members at a given time.

In the last chapter, I pointed out that even rocks are not as simple as we might imagine. Still, there is at least one way rocks are far simpler than groups: a rock is very closely tied to its constituting material. Even though the constituting material for a rock can change, we cannot get rid of it entirely and keep the rock in existence. If we get rid of the stuff that constitutes the rock, we have gotten rid of the rock. Groups are different. Not only can the membership of a group turn over again and again, but it is plausible that a group can persist even while it has no members at all.

Think, for instance, about the Supreme Court. Every one of us gets frustrated, from time to time, with their decisions. You have probably wished, as I have, that we could just clear out the court, somehow induce every justice to resign, so that we could fill the court afresh with different justices. This is unlikely to happen. But if it did, the Supreme Court would persist through the time it had no members, into the appointment of the new ones. We might even do things to the Supreme Court, such as rescind some of its powers, while it was memberless. The distinguishing characteristic of groups is that they are constituted by sets of people, when they have a constitution at all. But that does not mean that they always have to be constituted.[1]

How, then, is it possible to track a group over time? At a time the group exists, it seems there may be no "thing" to track. Is a group some kind of abstract object? If so, then how can it be constituted by people, as a rock is by a lump of minerals?

This seems like a deep, thorny question in metaphysics. Are groups concrete objects, or are they abstract objects? If they are abstract, how can they be constituted by people? If they are concrete, how can they exist without members? But this tangle is merely the product of a mistaken mindset about objects. It is hard to tear oneself away from the view that an object *really* is its material. A rock really is its minerals. A group really is its members. So when the group has no members, there is no object at all.

I want to shift this mindset by considering more generally how we identify objects over time. The view that a group is just its members is tempting, in large part because it seems hard to identify a group at a given time except by pointing to its members. But it turns out to be easy to identify objects in many ways, not just by their constitution. And it is easy to identify groups, even without their memberships.

To show this, I will modify a venerable tool of metaphysics: the "criterion of identity." Criteria of identity are widely used, but are also widely misunderstood. And they have long been construed much more narrowly than they need to be. Fortunately, we can easily generalize them, into what I call "cross-identifying criteria." With this more powerful tool, we can see that group memberships are just one of many ways to designate and track groups over time and possibilities.

Talking about Identity

To address the identity of groups, I will begin with a few concepts developed in the literature on the identity of *persons*.

Suppose one night Bob gets very, very drunk, and robs a convenience store. He wakes up the next morning and genuinely does not remember anything that happened the night before. Is Bob responsible for the robbery? John Locke, in

---

[1] Putting the point so starkly might rekindle a reaction I flagged earlier: that the Supreme Court is not a group, but rather something like an "organization." And therefore, that I interpret 'group' too broadly. I take this up in chapters 17 and 18.

chapter 27 of his *Essay on Human Understanding*, argues that he is not. For Bob to be responsible, he has to be the person who robbed the store. According to Locke, he is not that person: personal identity, in Locke's view, requires the sharing of memories. For the person who robbed the store to be the same person who woke up with a hangover, the memories of the first person have to overlap significantly with the memories of the second person. This means that the person who robbed the convenience store was not Bob. The robber might not even have been a person, if at the time of the robbery he did not satisfy whatever psychological requirements there are for being a person. But even if he was a person, according to Locke the robber was a *different* person from the one who woke up with a hangover.

Locke's memory theory is intended to give a set of conditions, or a *criterion*, for the identity of persons. Putting it roughly, a criterion of identity for persons answers the question *What makes two things the same person?* But this is a little confusing; after all, it almost seems illogical to say: *take two things, now find the conditions for them to be the same thing.*

This confusion goes away, though, if we formulate criteria of identity more precisely. A typical way of thinking about personal identity, and about the persistence of objects in general, is using relations between "stages" or "temporal parts" of the objects. A stage can be thought of as a momentary chunk of material—a three-dimensional snapshot of the person robbing the convenience store at 03:00:00.00 sharp, for instance, or a three-dimensional snapshot of the person hung-over at 10:00:00.00.[2] Call the 3 a.m. stage a and the 10 a.m. stage b. These stages, a and b, can each be seen as temporal parts of a person, perhaps parts of one person who is extended over time, or perhaps parts of two people. There is the person of whom a is a stage at 3 a.m., and there is a person of whom b is a stage at 10 a.m. This allows us to make perfect sense of the question whether these are the same person.

Here let me flag a point I will return to shortly. Talk of "stages" will not quite do, when we consider groups. If a group can persist even while it has no membership, then there will be times when it has no "stage." So we will need to generalize this typical way of thinking about criteria of identity, in order to apply it to groups. But first I will take a minute to clarify the standard approach, and after that will go on to generalize it.

A common formulation of a criterion of identity for persons is this:

(12.1) For all x and y, the person of whom x is a stage is the same as the person of whom y is a stage if and only if x and y stand in relation R.

---

[2] This is the way a four-dimensionalist understands stages. One way for a three-dimensionalist to understand a stage is as a pair of a person and a time. (See Sider 1999; Ninan 2009.)

A formula like this gives the criterion of identity for a kind of thing, such as *persons*, in terms of some other kind of thing, such as *stages*. The relation R is called a *criterial relation* for the kind. The idea is that if the criterial relation R holds between two stages, a and b, then that is enough to guarantee that a and b are both stages of the same person. R is the minimum needed to guarantee this: the idea of R is to give only the fewest requirements we need to guarantee that the person a is a stage of is the same as the person b is a stage of. In the formula, the reverse holds as well: if a and b are stages of the same person, that guarantees that they stand in relation R.

Notice that in this formula, the stages serve two distinct roles. One role is to be designators—that is, to designate *the unique person of whom x is a stage* and *the unique person of whom y is a stage*. The other role is to be the relata of R. That is, to be the things that must satisfy various conditions, in order to guarantee that those designated people are the same person. The formula gives a condition under which the two things designated—*the person of whom x is a stage* and *the person of whom y is a stage*—are identical. And that condition is that x and y stand in relation R.

Filling Out the Formula

Formula (12.1) is a common way of expressing a criterion of identity. However, it is elliptical in a way that can be confusing. The problem is this: people usually intend the criterial relation R to perform one task, but as the formula is written, there are three tasks implicitly packed into it.

In order for x and y to serve as designators, each has to pick out a unique person. There has to be one unique person of whom x is a stage, and one unique person of whom y is a stage. This means that unless we add something to the formula, it has to be the job of the criterial relation R to guarantee this. R has to guarantee that x is a stage of a unique person, *and* R has to guarantee that y is a stage of a unique person, *and* R has to guarantee that those uniquely designated people are identical. But we only want R to do the last of these. The job of R should be that *if* the stages uniquely designate people, then their standing in relation R guarantees that the designated people are the same person. If they do *not* uniquely designate people, then it is irrelevant whether or not the criterial relation holds of them.

For instance, suppose I get bitten by a zombie on March 15, 2030. Which of course, will turn me into a zombie. Now consider two stages: one before and one after. Stage s_1 is the stage at 10 a.m. on March 14, before the bite. And

s_2 is the stage at 10 a.m. on March 16, after the unfortunate transformation. s_1 is a stage of a person, but s_2 is not: it is a stage of a zombie. Therefore, it is irrelevant whether R holds between s_1 and s_2. Even if R does hold between those stages, that does not guarantee that s_1 and s_2 are stages of the same person. They are not, since s_2 is not a stage of a person at all. However, that is not a strike against R: the case does not show that R is an inadequate criterion of identity for persons. The job of R is to guarantee identity only for stages of unique people. If one of the stages is not a stage of a unique person, then R is off the hook.

To make this clear, we should reformulate (12.1) to explicitly limit the burden on R:[3]

(12.2) For all objects x and y, *if* there is one unique person of whom x is a stage and one unique person of whom y is a stage, *then* those people are identical if and only if x and y stand in relation R.

Let's introduce some symbolism, to write out (12.2) even more explicitly. (This is helpful for understanding what a criterion of identity does, and also for generalizing it, as I will shortly.) Let us introduce the following letters, and for the present case, the following interpretations:

L: the property *being a stage*

K: the property *being a person*

D: the relation *is a stage of*

LKD: the property *being a stage of some person*

LKD: the property *being a stage of exactly one person*

The properties LKD and **LKD** are built out of three parts: the property *being a stage*, the property *being a person*, and the relation *is a stage of* between them. With these, we can rewrite (12.2) more clearly, as:

---

[3] Sometimes this is done by restricting the domain of the quantifiers: that is, rather than "for all objects x and y...," saying instead, "for all stages x and y..." The problem is that this restriction is inadequate, because x being a stage does not imply that x is a stage of a person, and certainly not that x is a stage of a unique person. So if (as we usually do) we want the criterial relation to do just one thing, then we would need to restrict the quantifiers to "for all stages of a unique person x and y." But this formulation is confusing. It is better just to make everything explicit inside the formula.

(12.3) For all objects x and y,

if: (x is **LKD** and y is **LKD**),

then: [(the person z such that xDz is identical to the person v such that yDv) if and only if xRy.][4]

It is worth being so explicit: philosophers, in speaking of criteria of identity, frequently focus on only R. And in doing so, they frequently state the formula incorrectly.[5] If we are trying to be explicit about the conditions for two stages to be stages of the same person, we do not only have to ensure that they stand in R. We *also* have to ensure that the stages are person-stages, *and* that they designate a unique person. *All* those conditions on stages need to be worked out, to ensure that we are identifying a single person with the two stages.[6] And the analysis of properties L, K, and D are often just as revealing about "the nature of persons" as is the criterial relation R.

Altogether, in a criterion of identity, we take a pair of objects, x and y, and these objects play two distinct roles. First, we use them as designators. Each designates a unique object of kind K. Second, we have the criterial relation R. R is relevant only when x and y both do their designating jobs. When they do, R is a necessary and sufficient condition for x and y to designate the same object.

Other Ways of Designating

Criteria of identity are so commonly written in terms of relations between stages that one may easily get the misimpression that that is the only way they can work. Moreover, philosophers sometimes say that a criterion of identity *explains* why two stages are stages of the same person, or even *grounds* the fact that two stages are stages of the same person.[7] Sometimes a criterion of identity is believed to tell us "what it is" to be a person. This overestimates what a criterion of identity does. Criteria of identity simply give us one piece of useful

---

[4] In logical notation: $\forall x,y((\mathbf{LKD}x \wedge \mathbf{LKD}y) \rightarrow [(\iota z\mathrm{D}xz = \iota v\mathrm{D}yv) \leftrightarrow \mathrm{R}xy])$, where ι is Russell's iota (definite description) operator. (In the main text, I use infix notation for relation D, for readability.)

[5] See note 3 above.

[6] Some questions about personal identity can be answered without all these conditions. Lewis 1976, for instance, addresses the question of survival. But his R- and I-relations alone are not enough to guarantee identity.

[7] See, for instance, Hale 1987, 59; Lowe 1997, 628; 1998, 28–57; 2009, 18–19, 22–24; Williamson 2013, 148–52.

information about objects of a given kind. And stages are far from the only way we can designate objects. It is often more convenient to formulate criteria of identity that relate things other than stages.

For example, one way of designating people is by the laptop computers they own. (The fact that not everyone owns a laptop just means that this criterion will not cover all people.) We could write a criterion for personal identity in terms of laptop computers and the *owns* relation, rather than stages and the *is a stage of* relation. We can use the same symbols for these different designating kinds. Interpret the symbols as follows:

L: the property *being a laptop computer*

K: the property *being a person*

D: the relation *is owned by*

LKD: the property *being a laptop owned by some person*

LKD: the property *being a laptop owned by exactly one person*

With this, we can express the criterion of personal identity in the same way as above, with formula (12.3). Filling in the new interpretations of the symbols, we get:

(12.4) For all objects x and y,

if: (x is a laptop owned by exactly one person, and y is a laptop owned by exactly one person),

then: [(the person z such that x is owned by z is identical to the person v such that y is owned by v) if and only if xRy].

In this case too, relation R gives a minimal sufficient condition for personal identity. Only this time, it gives conditions on two laptops, each of which is owned by exactly one person.

Let's generalize this point. We start with objects of some kind L, such as *stages* or *laptops*. Those objects, together with some relation D (such as *is a stage of* or *is owned by*) are used to designate unique objects of some kind K, such as *persons*. The kinds K and L can be just about anything we want, and so too the relation D between them. We could for instance, have a criterion for the identity of persons in terms of laptop computers and the *was used to write the résumé of* relation, or in terms of cups of coffee and the *was the first thing drunk in the morning by* relation, or whatever else you like. There is nothing special about stages, and nothing deep about criteria of identity. All the criteria do is give us a piece of information about how to guarantee the identity of some designated things, in terms of some designating object and designating relation.

The Role of Time in Criteria of Identity

We will return to groups in a moment, but let's keep at this. We need to generalize criteria of identity still more, to incorporate time. Remember that different sets of people can constitute a group at different times. So if we are going to use memberships to designate groups, we have to do so *at a time*.

The literature on identity focuses almost exclusively on stages. But stages are unusual: a given stage only exists at a moment in time. So theories that only identify people using stages do not have to add a parameter for time. (If x is a stage of z, then it adds nothing to say that x is a stage of z at time t.) However, we want to use other identifiers, such as sets of people. So we need to generalize the standard formula for criteria of identity to include time:

(12.5) For all objects x and y, and for all times t_1 and t_2,

if: (x is **LKD** at t_1 and y is **LKD** at t_2),

then: [(the object z such that xDz at t_1 is identical to the object v such that yDv at t_2) if and only if R holds between x-at-t_1 and y-at-t_2].[8]

We can understand each of the properties to hold relative to a time. The criterial relation R we look for is one that holds between x at t_1 and y at t_2.

With this apparatus in hand, we can make sense of the question I began the chapter with. We were wondering whether the team your roommate is playing against tonight is the same team as the one they played before. In this example, we look at two sets of people—one set of opposing players tonight, and one set of opposing players a month ago. And we want to see if these two sets satisfy the condition for *constituting* the same team at the relevant times. To work this out, we interpret the symbols as follows:

L: the property *being a set of people*

K: the property *being an intramural basketball team*

---

[8] In logical notation:

$\forall x,y \forall t_1,t_2((\mathbf{LKD}x \wedge \mathbf{LKD}y) \rightarrow [(\iota zD(x,z,t_1) = \iota vD(y,v,t_2)) \leftrightarrow R(x,t_1,y,t_2)])$.

The domain of the second quantifier is restricted to times. Unlike restrictions on the first quantifier, this does not create any confusion.

D: the relation *is constituted by*

LKD: the property *being a set of people that constitutes some intramural basketball team*

LKD: the property *being a set of people that constitutes exactly one intramural basketball team*

This makes formula (12.5) say the following:

(12.6) For all objects x and y, and for all times t_1 and t_2,

> if: (x is a set of people that constitutes a unique team at t_1 and y is a set of people that constitutes a unique team at t_2),
>
> then: [(the team z such that x constitutes z at t_1 is identical to the team v such that y constitutes v at t_2) if and only if R holds between x-at-t_1 and y-at-t_2].

In order to track teams, using their sets of constituting players, this lays out the questions we need to work out:

- What is property LKD: under what conditions does a given set of people constitute some team at a time?
- What is property **LKD**: take a set of people that constitutes a team. Under what conditions does it constitute a single *unique* team at a time?
- What is relation R: under what conditions is the team uniquely constituted by one set of people at one time identical to the team uniquely constituted by another set of people at a different time?

We do not even have to use stages (or constituting sets) of a group as identifiers or as the relata of R. This answers the puzzle we raised at the outset: given that groups do not have to be constituted at all times when they exist, how can we identify teams before and after they are constituted? We simply designate them with *other* designators, not with their constituting sets.

Groups can be designated in various ways. Criteria of identity for groups can be written in terms of any other object that we can use to designate groups. We could give the criteria of identity for a team in terms of the person who manages it, or in terms of the corporation that owns it. Or we could give the criteria of identity for the Supreme Court in terms of cases it hears. Or in terms of events that bear some relation to the Supreme Court.

I will illustrate all this with an example, but first will add one more crucial amendment to the standard formula for criteria of identity.

Cross-Identifying Criteria

A criterion of identity is simply a set of conditions for guaranteeing identity of a member of the kind, where two objects play dual roles—they identify the thing, and then they are related by the criterial relation. It focuses our attention on the conditions for ensuring that we are identifying one thing, when we designate it with two different designators.

Understanding criteria of identity in this way reveals a historical quirk, which has imposed an unnecessary restriction on them. John Locke, Gottlob Frege, and the many other users of identity criteria have always taken them to relate two things *of the very same kind*. Two stages, for instance. Often, however, we want to ask questions about an object not designated by two objects of the same kind L (such as two stages, two laptops, two court cases, etc.), but by two objects of *different* kinds.

At the outset, I asked about identifying a team by the set of people playing now and by the set playing earlier. But we could ask a related question that uses two different identifying objects. For instance, we could ask whether the team playing now is the same one your roommate stayed up all night creating on IMLeagues.com. Or we could ask if the band coming out of the tour bus is the same one that wrote the song you just heard on the radio. For practical purposes, we are often interested in the identity of objects designated in two different ways.

There is no reason to rule this out. We can designate an object in one way at one time, and designate an object in a different way at a different time. And then give a relation between those different designat*ors* that guarantees that the designat*ed* objects are identical. This has the potential to vastly expand the utility of criteria of identity.

Here is a question we might ask about groups: *Why does {Holmes, Brandeis, Taft, Stone, Devanter, McReynolds, Sutherland, Butler, Sanford} constitute the Supreme Court in 1929 and {Alito, Breyer, Ginsburg, Kagan, Kennedy, Roberts, Scalia, Sotomayor, Thomas} constitute it in 2014?* An old-school criterion of identity would answer this in terms of the relation between the sets. But that is not the most useful explanation. A better explanation is that both sets, at their respective times, satisfy the conditions I described in the last chapter.[9] Sure, we could describe the set constituting the Supreme Court in 2014 in terms of appropriate additions and deletions stretching back to 1929. But the more basic explanation is between a stage and something about the original court

---

[9] That is: that the Supreme Court exists and that the set is the largest set of people each of whom has Q at that time.

at its formation, not between two stages. This mixed pair of designators—an originating event and a stage—will turn out to be useful.

Let us modify the more traditional form of criteria of identity into what I will call a *cross-identifying criterion*. A cross-identifying criterion is not just about two kinds, an identified kind K such as *person* and a pair of identifiers of a different kind L, such as *stages*. Rather, it is about three kinds: an identified kind K and then two potentially different kinds of identifiers L_1 and L_2. For instance, we might have a criterion of identity for persons, using stages and the *is a stage of* relation as one designator, and laptops and the *is owned by* relation as the other designator.

Figure 12A depicts it visually.

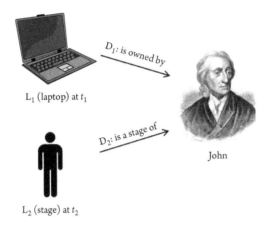

Figure 12A Cross-identifying an object, with objects of two different kinds

A cross-identifying criterion gives the conditions under which the unique thing designated in one way at one time is guaranteed to be identical to the unique thing identified in a different way at a different time.

With this, we can make sense of the conditions under which the team playing now is the same as the team your roommate spent all night creating. For this case, we will distinguish two different designating kinds L_1 and L_2, and two designating relations D_1 and D_2:

L_1: the property *being an event*

D_1: the relation *initiated*

L_2: the property *being a set of people*

D_2: the relation *is constituted by*

K: the property *being an intramural basketball team*

LKD₁: the property *being an event that initiated exactly one intramural basketball team*

LKD₂: the property *being a set of people that constitutes exactly one intramural basketball team*

Our final formula—the cross-identifying criterion—is only a slight generalization of what we have already written above:

(12.7) For all objects x and y, and for all times t_1 and t_2,

if: (x is **LKD₁** at t_1 and y is **LKD₂** at t_2),

then: [(the object z such that xD_1z at t_1 is identical to the object v such that yD_2v at t_2) if and only if R holds between x-at-t_1 and y-at-t_2].[10]

This formula says that R holds between x at t_1 and y at t_2 just in case the unique team initiated by x at t_1 is identical to the unique team that y constitutes at t_2 (supposing that there is a unique team initiated by x at t_1 and that there is a unique team that y constitutes at t_2).

The cross-identifying criterion is a replacement for the traditional formulations of criteria of identity. I will be applying it to groups, but it is a general tool for metaphysics. All the preceding formulas are just special cases of this cross-identifying formula. To go back to the preceding formulas, we can make one or both of the designating relations timeless. To do that, we just make relation D_1 or D_2 (or both) hold at all times. We also can have L_1 and L_2 be the same kind, and have the D_1 and D_2 be the same designating relation. So the formula used in (12.3) is just a special case of (12.7): it is a version that is timeless and that has just one designating kind.

And it makes no sense to speak of "the" criterial relation for a given kind, like for persons or for groups. For any kind K, there is a family of criterial relations, each one applying to a particular choice of designator kinds L_1 and L_2 and designating relations D_1 and D_2.

Persisting without Material

Getting clear on criteria of identity, and on how they apply to groups, accomplishes two things. As we will see in the next chapter, thinking about cross-identification

---

[10] In logical notation:

$\forall x,y \, \forall t_1,t_2 \, ((\mathbf{LKD_1}x \wedge \mathbf{LKD_2}y) \rightarrow [(\imath z D_1(x,z,t_1) = \imath v D_2(y,v,t_2)) \leftrightarrow R(x,t_1,y,t_2)])$.

is a useful tool for working through frame principles for groups of certain kinds. More importantly, however, it helps to shift our mindset away from the idea that objects we encounter in the everyday social world are just the material that constitutes them. This shift is not easy to do. Even when we see that a group is distinct from the set of people constituting it, it is hard to let go of the notion that a group *really is* the constituting set.

Having generalized criteria of identity, we can see that even if we want to identify a group at any time that it exists, we do not need stages at all those times. In fact, the stages of a group are just one of its features. There is, of course, a close tie between a group's constituting set and the group itself: what qualifies groups as material objects—just as with statues and people—is precisely that they are materially constituted. Groups just also have the feature that they do not *always* have to be materially constituted. But other than that, the constitution of a group is about as closely tied to the group as the constitution of a statue or of a person is tied to it. Tightly related in some ways, and not so much in others.

In this chapter, I have explained how we can identify groups even when they are empty. But the point is not really about the possibility of empty groups; that does not matter much. Instead, the aim is to use that example as an instrument for rethinking the ways we track groups over time and across possibilities, and for showing that a group's constituting set is just one among many features of the group.

One last point. In writing out formulas for criteria of identity, I have not used the grounding relation. There is good reason for this: *criteria of identity do not give the grounds for facts about persistence or identity*. The fact that I am identical to myself is not grounded by the fact that two stages of me stand in a criterial relation to one another. Nor is it grounded by the fact that two laptops I own stand in a criterial relation to one another.[11] Presumably, the fact that I am self-identical is grounded by the fact that I exist, nothing more.[12]

Instead, criteria of identity can be understood to be convenient systematizations of frame principles. They are consequences of frame principles for the existence of a group of some kind, together with frame principles for the constitution of a group of some kind. When, in the next chapter, I work out some sample frame principles like these, it becomes straightforward to see how criteria of identity—both standard and cross-identifying ones—can be derived.

---

[11] A failure to recognize this is the source of the fruitless debate between Lowe and Williamson over which is more "fundamental," the one-level criteria of identity or the two-level criteria of identity. See Lowe 1991; Williamson 1991, 2013.

[12] There is nothing special about people in this regard. Regardless of the object x, it does not take much to ground the fact $x=x$.

13

Kinds of Groups

Even in this abstract work on the metaphysics of groups, practical social science is always at the back of our minds. How can we reduce legislative corruption? Improve the performance of high schools? Make corporate boards more effective? These are not questions about *particular* groups, like the Supreme Court. Rather, each is a question about groups of some *kind*.

Examples of kinds of groups include *local courts, circuit courts, district courts,* and just *courts*. The US Senate is a particular group, but *senate* is a kind of group, of which the US Senate is an instance. Other kinds of groups are *tax bureaucracies, administrative bureaucracies, legislatures, dockworkers' unions, boy scout troops, men's clubs, treehouse clubs, corporate boards,* and so on. In the social sciences, kinds of groups play a more central role than do particular groups. When we build models to address questions like the ones above, we build them atop assumptions about the grounding of facts about legislatures, facts about high school classes, or facts about corporate boards. Model-building implicitly takes a stand on the grounding of facts about groups of a given kind or other.[1]

Unfortunately, when it comes to kinds, it is easy to get grounds drastically wrong. It might seem that moving to kinds would simplify things. After all, facts about senates are more generic than facts about the US Senate, and facts about circuit courts are more generic than facts about the 9th Circuit Court of Appeals. It is often assumed we can treat groups of a given kind, like senates or circuit courts, as generic sets of slots or roles.[2]

---

[1] It is important to distinguish facts about kinds of groups from facts about groups of a given kind. This sounds hair-splitting, but is not. Compare, for instance, the fact **The kind legislature exists** and the fact **A group of the kind legislature exists**. These are two extremely different sorts of fact. In this discussion I am considering facts like the existence of legislatures (i.e., groups of the kind *legislature*), not the existence of kinds.

[2] This thinking is often baked into the way groups are represented in typical models. Commonly, a kind of group is represented as a structure consisting of an array of sets of attributes, with each element of the array representing a slot for a member of the group.

In chapter 11, I pointed out the attributes of group members are often not enough to ground even simple facts about a particular group. Even the simplest facts about the Supreme Court depend on much more. When we turn to kinds, the problem gets worse. Facts about groups of a given kind involve at least as diverse a set of grounding conditions as do the facts I discussed about the Supreme Court. And many such facts have little to do with the group members at all. If we represent a kind of group as a generic array of slots and rules, we capture very little about what matters even for general facts about groups of that kind.

Here I will focus on a new example. The Supreme Court was an unusual and carefully selected case, a conspicuous group whose frame principles are largely codified in the written law. We live our everyday lives, however, amid informal and unremarkable groups. The Supreme Court spared us the effort of analyzing it from scratch. But too much attention to formal groups can give the mistaken impression that most groups are intellectual creations, whose properties we are explicitly aware of. So now I want to turn to a more commonplace example. An example like the one I mentioned in the last chapter: the kind *intramural basketball team*.

I should forewarn the reader that, of all the chapters in the book, this one gets into the most nitty-gritty detail. To some extent, that is the point: I want to work through one example where we treat the grounding inquiry with care. It is impossible to give a mechanical algorithm for churning out frame principles. The best way to clarify the inquiry is to reason our way through a detailed example. The key aims of the chapter, then, are to show how the grounding inquiry is done, for the sorts of facts we deal with every day in the social sciences, and that their grounding conditions involve much more than generic facts about slots or roles or group members. Then, as a side-benefit, to show how we can derive criteria of identity from such frame principles.

Intramural Basketball Teams

Even among sports teams, there are countless kinds. In a typical American university, there are kinds of varsity teams, club teams, intramural teams and a great many kinds of "pick-up" teams. All these kinds of teams have different constitution and existence conditions from one another. Varsity teams typically have formal tryouts, selection processes, and authority figures who come to agreement with the players about their joining the team. Club teams also compete with teams from other schools, and they also tend to have formal hierarchies and rosters. Intramural teams also have rosters, but are less formal about who plays what position, and only play other teams in the same school. Pick-up teams are organized on the spot.

I choose intramural basketball teams partly because they are mundane, and partly because they occupy an interesting middle ground. Intramural teams are not formalized like the Supreme Court, nor are they spontaneous and self-organized like pick-up teams. Their rules are not enshrined in law books, but they nonetheless have a certain structure, just enough to accomplish practical aims. To be sure, intramural basketball teams are an unimportant kind of group. Their ordinariness is a virtue: the details we root out are representative of the sorts of details we might find for many other run-of-the-mill cases.

Getting Started on Existence

At Tufts, there is a rough procedure for creating and managing intramural teams, much of which has been outsourced to the website IMLeagues.com. A new intramural team is created in a sequence of events. It is initiated by someone (usually a student) who has registered as a "manager." The steps could be broken out in limitless detail, but basically, they are: (1) The manager goes to the relevant website and clicks to create a team within an existing league. (2) A form is sent from the computers at IMLeagues.com to the manager's computer. (3) The manager fills in the relevant fields on the form, and clicks to submit it. (4) The database at IMLeagues.com is populated with the relevant information. (5) An email is automatically sent to the relevant supervisor at the athletic department. (6) The supervisor then clicks, and so on, and fills in the forms to approve the team. (7) The database at IMLeagues.com is populated with the relevant information. Once these steps are performed, a roster is generated for players to join the team.

Just because the team is created in this sequence of events, however, does not mean that all these steps are *grounds* or *metaphysical reasons* for the existence of the team. Many of the steps merely *cause* some grounding fact to be in place. The performance of steps (1) and (2), for instance, cause the registration form to be displayed. But these steps are just causes along the way to the obtaining of the facts that do ground the team's existence.

In working out grounding conditions, it is crucial to be sensitive to the difference between grounds and mere causes. It is not always obvious how to disentangle these. Some of the most intuitive lines we might draw between them do not work. One might think, for instance, that grounds are synchronic and causes are diachronic. (That is, that if G grounds F, then G and F are facts about events that occur simultaneously, whereas if C causes E, then C precedes E.) But this is not an adequate dividing line. It is very common for grounds to involve the occurrence of events over time. What grounds a fact like **An intramural basketball team exists** can be a process that takes place over time: among the grounds for a team to exist at time *t* are the occurrences of events at times prior to *t*. It is also important to recognize that some causal facts can themselves be grounds for other facts. For instance, a plausible ground for the existence of a team is a causal fact like **The manager caused the database to be populated**.

In short, it can be a delicate matter to identify the grounding conditions for a given fact. Still, we can come up with reasonable hypotheses, dividing the metaphysical explanations from the mere causes. Here is a stab at what matters for the grounding of the creation of an intramural basketball team: (1) facts about the physical record being populated in a particular way; (2) facts about the manager being the cause of the physical record to be appropriately populated; and (3) facts about the team's approval by the athletic department.

Supposing these are the grounds for a team to come to exist, they are compatible with a team existing even before it has any members. But whether or not a team can exist without members, one thing is clear about teams, which was muddy for the Supreme Court: teams do not exist necessarily. The existence of the Supreme Court might or might not be anchored in our frame.[3] But the existence of an intramural basketball team depends on some set of contingent facts obtaining.

Once an intramural basketball team is created, it does not require any ongoing maintenance. However, there are conditions for it to be removed or eliminated as a team. Teams last for only one semester, expiring on the last day of the term. In addition, intramural teams at Tufts expire automatically if a game is forfeited, which occurs if enough players do not show up on the field at the weekly time.

This structure resembles the one I introduced in connection with *membership* in the Supreme Court: an initiation, maintenance, and exit structure. Here, the three-part structure applies not just to membership in the group, but to the existence of the group itself. The structure is commonplace for the existence conditions of groups. Academic classes, corporate boards, and courts, for instance, also exist in virtue of some historical sequence of events occurring, and have various maintenance and exit conditions as well.

This is a pattern that continually crops up in social ontology: social facts commonly have diachronic grounds. Often, some effort and special circumstances are required to make a social entity exist, or for some property to hold

---

[3] See alternatives (1) and (2) regarding the existence of the Supreme Court, in chapter 11.

of it. And often, there are only minimal requirements for it to be maintained, and there is a regular expiry date or some kind of neglect, violation, or particular change in circumstances in order for it to be eliminated. (The same sort of structure even applies to the introduction of new words into a language. See Epstein 2008b.)

With the example in place, let us turn to the grounds of some simple facts about intramural basketball teams.

Existence

Here are a few facts about the existence of intramural basketball teams:

(13.1) *One new intramural basketball team comes to exist at time t.*

(13.2) *An intramural basketball team exists at time t.*

(13.3) *Intramural basketball team b exists at time t.*

Any of these might be relevant to some practical end. If we are interested in modeling the formation of new teams, we need to apply frame principles for facts like (13.1). If we are interested in modeling the number of teams around at any given time, we need to apply frame principles for facts like (13.2). And if we are interested in modeling facts about some particular team, or tracking a team over time or possibilities, we need to apply frame principles for facts like (13.3).

Initiating an intramural basketball team is not quite the same as initiating a one-off group like the Supreme Court. There is only one Supreme Court. If it had been established in 1795 rather than 1789, it would still be the Supreme Court. And it can only be initiated once. We could go through the procedure for establishing the Supreme Court repeatedly, but only the first time counts. In contrast, there can be many intramural basketball teams. Going through the initiation procedures once creates one new team, but going through it multiple times creates multiple teams.[4]

This means that we can write a very simple frame principle for the determination of (13.1). Whenever an initiation occurs, one new team comes to exist:

(13.4) It is necessary* that: if some event *e* occurs, ending at time *t*, which satisfies the initiation conditions for intramural

---

[4] Here we need to specify the initiation conditions carefully, to avoid taking what is really just one single initiation event to have the effect of creating more than one team.

basketball teams, then the occurrence of *e* grounds the fact that one new intramural basketball team comes to exist at *t*.

This frame principle is simple, because it does not require any machinery for tracking a team over time. Even so, it already shows that a generic fact about groups of a given kind need not have individualistic grounds. Nothing restricts what gets included in the initiation conditions for a team: they can involve more or less any grounding condition. There is no reason, that is, that the grounding conditions for (13.1) should be restricted to the individualistic ones.

It is with (13.2), however, that the fun begins: it is surprisingly complex to work out a frame principle for this fact. To ground it, we do not just need a team to be initiated. We also need to take into account that teams can be eliminated, between the time of initiation and time *t*. For fact (13.2) to obtain, at least one team needs to exist at *t*. This means that initiation has to have occurred at least once, prior to *t*, and also that at least one of the initiated teams still survives at *t*. A first pass at a frame principle, then, is:

(13.5) It is necessary* that: if there exists a team *x* such that *x* was initiated prior to time *t*, and such that *x* was not eliminated between the time of its initiation and time *t*, then that grounds the fact that an intramural basketball team exists at *t*.

In a sense, this frame principle gives a complete set of grounding conditions for (13.2). However, (13.5) is not particularly informative. The implicit point of the "grounding inquiry" we are working on has been to give the grounding conditions for facts about groups in terms of more basic facts.

The problem with (13.5) is that it gives two different properties one team must have, but it does not give the grounds for both properties holding of the *same* team. We know what it takes to initiate a team. And we know what an elimination event is for some team—either the semester ends, or else the team forfeits a game. But suppose you have two initiation events, in which, say, the "Red Team" and the "Blue Team" are created. And then you have one elimination event. Which team does the elimination event apply to? Effectively, this is a cross-identification problem, as I discussed in the last chapter. It is not mysterious why a particular forfeit is a forfeit by the Red Team, not by the Blue Team. But if we just say, "that later event is a Red Team forfeit," then we have not given those more basic grounds.

This is not just a technical quirk. When we model facts about groups of a given kind over time, or across different possibilities, we need to *track* them. Models often overlook this need. Even to ground generic facts like (13.2), we need to work out the grounds for the group initiated at a given time to have some property later on. We need the grounds for *that very group* to have the property, not just for *some* group to have that property. Just because we are interested in modeling facts about groups of a given kind in general, rather than facts about particular groups, does not mean we relinquish the obligation to track groups over time and over possibilities. This means that even a generic fact about groups of a given kind can depend on worldly and impersonal facts, just as facts about particular groups do. And it can depend on such facts in rather complicated ways.

Grounding (13.2), Part One: A Puzzle

We should start by spelling (13.5) out, to be precise about the pieces we need to deal with. For a team to exist at t requires that some team was initiated before t, and not eliminated between its initiation time and t. (Call the initiation time t_0, and the elimination time t_{elim}.) In more detail:

(13.6) It is necessary* that: if a sequence of events e occurs ending at a time t_0 before t, such that
(1) *(some team was initiated)* e satisfies the initiation conditions for intramural basketball teams, and
(2) *(that team was not eliminated)* there is no time t_{elim} between t_0 and t such that the team initiated by e is eliminated at t_{elim},
then that grounds the fact that an intramural basketball team exists at t.

Clause (1) is straightforward. Clause (2), however, gives rise to a puzzle. A team is eliminated if the semester ends, or if the team forfeits a game. For a team to forfeit a game is for an insufficient number of players to show up at a game. That is, suppose that at the time of a game, the team is constituted by set S of people. If an insufficient subset of S comes to the game, then the team forfeits the game.

Among the conditions for a team to be eliminated, in other words, are facts about the set S that constitutes the team at that time. The fact about whether it is eliminated at time t depends partially on facts about the membership the team happens to have at a particular time, and partially on other facts, having nothing to do with the membership.

But here is the puzzle. To address the elimination conditions of the team, we need to look into the conditions for a team to be constituted by some set. We have not yet gone into these conditions. However, we do know at least one of them: that the team exists. One condition for a team to be eliminated—for it not to exist—is that it exists!

A side-note here. In preparing this book, I have worked through many frame principles for social facts. This is the only complex one I present in detail in the book. But I do want to mention that similar problems crop up again and again. As I will discuss in a second, they can be addressed. This puzzle is not a deep paradox, just a hurdle. But the prevalence of this sort of problem made me wonder why I never used to run into problems like this before, in building models. Was it that my models involved simpler facts, where these nuances do not come up? Was it that these problems, these interdependent grounding conditions, were irrelevant to my models? Or rather, was it that I just never noticed them, because I was making rash assumptions about how social facts are determined? This last alternative is the one I have come to believe. There are complications and paradoxes we are not even aware of, if we are cavalier about the metaphysics of groups. This sort of hurdle is one we *should* encounter, and frequently. But it does not come up if we start with a simplified picture of social facts.

The interdependence of existence and constitution seems paradoxical. But one way to make sense of it is to bring time into the equation. Clause (2) says: there is no time t_{elim} between t_0 and t such that the team initiated by e forfeited at t_{elim}.[5] To make sense of the clause, we need to work out what it means for the team created by e to forfeit at t_{elim}. One of the conditions for this is that, prior to t_{elim}, the team created by e exists. And for that team to exist prior to t_{elim}, it cannot have been eliminated at a still earlier time.

So here is how we can go about it. Start with a team from the moment of its creation, that is, from time t_0. For some team to come to exist at time t_0, there has to be some initiation event e, ending at t_0. Because the team created by e is created at t_0, it cannot be eliminated at t_0. Instead, consider what can happen at the next moment t_1. It can continue to exist, or it can be eliminated if the team forfeits a game at that time. For the forfeit to happen, (1) the manager has to have scheduled a game for the team at t_1, and (2) there has to be some set S of people that constitutes the team at t_1, and (3) an insufficient number of people in S show up to the game.

It is not difficult to give grounds for (1) and (3), so I will focus on (2).[6] What is it for set S to constitute at t_1 the team created by event e? Again, the team has to exist. But this is already covered, since it could not have been eliminated

---

[5] I will just ignore the semester-ending elimination, since it makes things wordy without adding much insight.

[6] Working out clause (1) will require elaboration on what the grounds are for a game to be scheduled.

between t_0 and t_1. So here we can proceed with an account that does not make circular reference to the team itself.

Grounding (13.2), Part Two: Causal Links and the Constitution of a Team

Suppose we have two intramural basketball teams, the Red Team and the Blue Team. And suppose that the Red Team is constituted by set S_1 at t, and the Blue Team is constituted by set S_2 at the same time t. What grounds these facts? What makes it the case that the Red Team is constituted by S_1, and that the Blue Team is not?

The two groups are different—one is initiated by e_1, and the other is initiated by e_2. Subsequent events, such as events in which they add people to their respective rosters, are connected both causally and noncausally to those distinct initiating events.

Consider how an intramural basketball team is created. Involved in that event is a particular manager, a particular database the manager populates, and a particular approval event. Once the team has been created, fields get added to the database—that very part of the database that the manager populated—for players to add their information. (We can call it a "signup sheet.")

To be a member of *that* team is to populate that very signup sheet, the one for that team, and not a different one. For person x to be a member of the team initiated by event e_1, x must sign up on a signup sheet that is causally connected in the appropriate way to e_1. It is the fact that x is signed on a sheet causally connected to e_1, rather than to e_2, that determines which team x is a member of. Suppose the Red Team's signup sheet is a red sheet of paper, and the Blue Team's is a blue sheet of paper. It is Joe's signature on the red sheet—a fact about an ink mark on a piece of paper, together with a fact about the causal connection between Joe and the ink mark—that grounds Joe being a member of the Red Team. Joe's membership is grounded by physical facts about the world, and by the occurrence of a sequence of events involving causal relationships between the manager's actions, the physical medium, and Joe's actions.

This observation highlights two central themes of the grounding inquiry. First, even in dealing with generic facts about groups of a given kind, we need to (and we can) track and cross-identify particular groups. To ground such facts, we cannot neglect the sorts of facts in virtue of which a particular group persists. Second, these grounds are not just facts about the members, or about anybody's attitudes. They are grounded by concrete causal facts about physical stuff in the world: facts involving whatever we choose to anchor as the grounding conditions. In the present case, they are partially grounded by facts about a piece of paper or about a physical computer system, somewhere off in a remote datacenter. Here we have a very generic fact about a run-of-the-mill group. But even for this, there are no limits to the sorts of facts that can figure into its grounds.

Back to the constitution of a given intramural basketball team. Once we have clarified the membership conditions, we can generalize them into the conditions for a set of people to constitute an entire team. Let $Q(e)$ be the property of satisfying the membership conditions for a particular intramural basketball team initiated by event e. Then we can write the determination frame principle for the fact **y constitutes the intramural basketball team initiated by e at t** as:

(13.7) It is necessary* that: for all e, y, t: if e is an initiation event for an intramural basketball team and the intramural basketball team initiated by e exists at time t, and y is the largest set of people each of whom satisfies $Q(e)$ at t, then those facts ground the fact that y constitutes the intramural basketball team initiated by e at t.

Notice two things about this frame principle. First, it designates the *constitution* of a team by relating it to the *initiation event* of that team.

Second, notice that this frame principle exhibits the same sort of puzzle we saw in (13.6), but reversed. Among the grounds for the team's constitution at t is the team's existence at t. But, of course, the team's existence at t depends on its not having been eliminated prior to t. That involves the team's not having forfeited a game, which in turn involves facts about its constitution.

Again, this can be dealt with by breaking down the course of events moment by moment. At t_1, the players have not yet had a chance to drop out of the team. So the constitution of the team is simply the set of people who have signed up on the sheet at time t_0. This is enough to give us what we need for existence at t_1. The team's constitution at t_1 then depends only on the facts about who has signed up in the appropriate way—that is, who has populated a database causally connected in the appropriate way to the database the team was initiated with. At t_1, we also have noncircular grounds for whether the team is eliminated—that is, whether an adequate subset of its constitution has failed to show up at a scheduled game.

Grounding (13.2), Part Three: Back to Existence

From here, we can finally assemble a full frame principle for (13.2). We have the conditions for a set S to constitute at t_1 the team created by event e. Using this, we can fill out the conditions for a team created by e to have forfeited a game at t_1: there must be a game scheduled for the team at t_1, and not enough people in S show up to it. For the team to *continue* to exist at t_1, then, those conditions must go unfulfilled.

At the next instant, t_2, there are two possibilities: either the team was eliminated at t_1, or else it continues to exist at t_2. If it continues to exist at t_2, then either the membership remains the same, or it changed at time t_1. Similarly at t_3, t_4, and so on.

To generalize, consider the conditions for the team created by event e to forfeit at time t_n. There are two conditions:

(1) The team initiated by e exists up to t_n: This means that there is no time t' between the team's initiation and t_n, such that the team initiated by e forfeited a game at t'
(2) The set constituting at t_n the team initiated by e fails to show up to one of that team's games at t_n.

Notice that condition (1) involves the team forfeiting a game at a time. But this condition is in terms of earlier times. So it can be unwrapped back to the beginning: we can "recursively" analyze what it is for a team to forfeit at t_n in terms of earlier times, and those in terms of still earlier times, all the way back to the initial creation of the team.

The same goes for condition (2). Spelling it out involves the fact that S constitutes at t_n the team created in e. This, in turn, can be understood as:

(2_a) The team initiated by e continues to exist up to t_n (This is the same as (1))
(2_b) The elements of S are the people who signed up on the sheet, minus those who dropped out, starting from the time of the initiation up to t_{n-1}.

Taking all this together, we get the conditions for a team initiated by e to forfeit a game at t. The grounds for the team's existence and constitution are interleaved with one another, from the time it is created until the time it is eliminated. And with this, we have written out a frame principle for (13.2) without referring to teams in the analysis. All the references to teams at time t_n are replaced with reference to teams at t_{n-1}, and those with references to teams at t_{n-2}, all the way back to the initiation event, which does not make reference to teams at all. Thus we have the full grounding conditions, in non-team-involving terms, for (13.2), by replacing the above recursive analysis into (13.5).

Out of the Weeds

That was a lot of work. But from it, we get several insights.

Despite the interdependence of the existence and constitution conditions, we were able to work out the grounding conditions for a social fact like (13.2) in terms of more basic facts. We might not always be able to give a kind of reductive set of grounding conditions for all social facts. Sometimes they may be too complicated, or too interwoven with other social facts, for a frame principle to be worked out. But here we see that the existence and constitution of a team over time both are grounded by facts about what we do, about systems in the world, about procedures we follow, and so on.

The grounds of a generic fact like (13.2) involve all the complex and worldly facts that are involved in grounding facts about a particular group. Just because we have moved from particular groups to kinds of groups does not mean that we can put aside specific facts about the world like the ones we talked about in connection with the Supreme Court. Key to this is that even a generic fact like (13.2) involves tracking particular groups. It involves the sorts of cross-identification I discussed in the last chapter. Even to conclude that a team exists at time t, we need the conditions for tracking a particular team-elimination event back to a particular team-creation event.

The frame principles even for simple facts about this informal kind of group are extremely detailed and textured. They are not likely to be things we have attitudes towards, at least until we work them out. Instead, they are largely anchored by facts about practical stuff in the world (such as computer systems), by causal facts about the world, and by a few poorly worded rules on IMLeagues.com.

The grounding conditions involve facts about the world, which the people involved may or may not know about. An intramural basketball team, for instance, can be eliminated just by satisfying the elimination conditions, whatever they are, even if no one knows about it. If, for instance, everyone at Tufts falls asleep for 48 hours straight, then several intramural basketball teams will have been eliminated because they have forfeited their games, even if no one is aware of it. Anchored as they are, those conditions involve *facts* about whether people have shown up, not their *beliefs* about whether they have. It can be tempting to think that we can only create or eliminate teams and other entities like these by having something in mind, or deciding it's eliminated, or maybe by all forgetting about them. But this example exhibits clear elimination conditions, which just involve some fact obtaining in the world.

Deriving Criteria of Identity

Out of these frame principles, we also get a nice result. We can easily derive a variety of standard and cross-identifying criteria of identity.

A standard criterion of identity is what we might call a "stage/stage" criterion. It gives a relation holding between two "stages." When each of those stages is a stage of exactly one group, then the relation holding between them is sufficient to guarantee the identity of the groups having them as stages.

When the criterial relation holds between two stages of an intramural basketball team, that is sufficient for them to be stages of the same group. But the criterial relation is not much of an *explanation*. The metaphysical reason that two stages are stages of the same intramural basketball team need not include anything the stages themselves have to do with one another. Instead, what matters is that the stages trace back in the appropriate way to the same initiating event. The stage/stage criterion can easily be derived from a more basic cross-identifying criterion—a stage/initiating-event criterion.

That criterion is more useful than the stage/stage criterion. But it too is derivable from more basic principles. Namely, from the frame principles for existence and constitution we just discussed.

To sketch the derivation, I will start with the cross-identifying formula I gave in the last chapter, with the same interpretations of the symbols:

L_1: the property *being an event*

D_1: the relation *initiated*

L_2: the property *being a set of people*

D_2: the relation *is constituted by*

K: the property *being an intramural basketball team*

LKD$_1$: the property *being an event that initiated exactly one intramural basketball team*

LKD$_2$: the property *being a set of people that constitutes exactly one intramural basketball team*

For convenience, here again is the formula for a cross-identifying criterion:

(12.7) For all objects x and y, and for all times t_1 and t_2,

if: (x is **LKD$_1$** at t_1 and y is **LKD$_2$** at t_2), then: [(the object z such that xD_1z at t_1 is identical to the object v such that yD_2v at t_2) if and only if R holds between x-at-t_1 and y-at-t_2].

The criterial relation R holds between x at t_1 and y at t_2 just in case the unique team initiated by x at t_1 is identical to the unique team of which y is a stage at t_2. So interpreted, (12.7) gives a cross-identifying criterion: a stage/initiating-event criterion for the identity of intramural basketball teams.

Using frame principles for existence and constitution of an intramural basketball team, we can derive all of the components of (12.7). Relation R, for instance, is simple. R needs to apply only if stage y uniquely designates a team at t_2. And that means that R needs to apply only if the team y designates *exists* at t_2. Which means that R does not need to include any of the back and forth about constitution and existence: we can take it for granted that the team exists at that time. All that R needs to include, then, is that the members of y satisfy the signup process, that is, that they be appropriately connected to the signup sheet that, in turn, is connected to event x.

Property **LKD₁** is also fairly simple. For an event to be the initiating event of an intramural basketball team is just for it to satisfy the initiation conditions for intramural basketball teams. For it to initiate a unique team requires nothing more, since we stipulated that one event creates exactly one team.

The only complicated property is **LKD₂**. What are the conditions on a set of people, for it to constitute an intramural basketball team at a given time? For this property to hold of a set, the set has to trace back to an initiating event in the appropriate way, and there cannot have been an elimination event for that team in the intervening time, between the initiation and t_2. Thus this property uses all the details of the interleaved frame principle I discussed above. It also includes an additional condition that there is no other team that has those characteristics. That is, there is no other tracing-back to an initiating event and elimination failure. This property, too, involves all the interleaving and a bit more. With the work we've done on the above frame principles, it would not take much more work to derive **LKD₂**.

That sketches, at least, how we can derive a stage/identifying-event criterion from the frame principles above. And once we have the stage/identifying-event criterion, it is entirely straightforward to derive a stage/stage criterion: two stages are stages of the same team just in case there is one event they both cross-identify with.

Moving On

In this chapter, I have turned a microscope to one generic fact about one mundane kind of group. If we want to model what makes the fact obtain, we first need a decent understanding of its grounding conditions. It is too easy to put blinders on, and not notice how diverse these grounding conditions are. And then to build faulty assumptions into our models. Before we build our models, we need to identify the facts we are modeling, and make an effort to work through their grounding conditions.

A good first step in modeling social facts is to understand what is being modeled. That is, to understand their grounding conditions. Until we see how complicated the grounding conditions can be, we may not even realize that this project deserves serious time from model builders.

In the next chapter I shift gears. I turn to the topics that have drawn by far the most attention in theories of groups: *group action* and *group intention*.

14

Group Attitudes: Patterns of Grounding

People have knowledge and beliefs, desires and fears, preferences and aversions. They plan, judge, form intentions, and take action. Individuals sometimes take pride in what they have done and sometimes regret it, and they bear responsibility for their actions. What about groups? Can a group form an intention, pass judgment, take action, or bear responsibility?

At one time, it was assumed that *all* social facts are facts about the "psychology of society." Some theorists, like Emile Durkheim, argued that group psychology is autonomous of individual psychology.[1] Others, like J. S. Mill and J. W. N. Watkins, held that it is nothing over and above facts about individual psychology.[2] But they tacitly agreed that social theory was a kind of psychology, even as they disagreed about the relation between individual minds and the "mind" of society.[3]

This assumption was tenacious, but began to lose its grip around the middle of the twentieth century. As I discussed in chapter 1, even Watkins eventually expanded the facts that count as "individualistic" to include nonpsychological facts. In the last chapters, I have argued this is not enough. Even when we restrict our attention to simple facts about groups, a great many of them are not grounded individualistically at all. We can balloon the notion of "individualistic," like Virchow ballooned the notion of "cellular." But unless we balloon it to the point that *every* nonsocial fact is artificially defined to be individualistic, then facts about individuals do not suffice to ground many social facts.

On the other hand, what about all the social facts that *do* seem to be about the "psychology of society" or the "psychology of groups"? What about group

---

[1] Durkheim (1895) 1982.

[2] Mill (1843–72) 1974, 879, 907ff.; Watkins 1953.

[3] I am only putting these points roughly. For a nuanced treatment of these and many more views, see Udehn 2001, 2002.

beliefs, intentions, and judgments? Not *all* social facts are psychological, but that does not mean that none are. Certainly we talk a lot about the attitudes and dispositions of groups. Here, for example, are some mentions of group intention from the newswire on a recent day:[4]

> **Parliament intends** that the Duke and Duchess of Cambridge's first child—regardless of gender—will succeed to the crown.[5]
>
> **A group of U.K. hardliners intends** to push the coalition government to make a "final gate" decision on moving forward with the replacement of the nation's Trident missile SSBN fleet prior to the 2015 general elections instead of afterward.[6]
>
> A person familiar with the Yankees' deliberations says **the team intends** to discipline Alex Rodriguez for seeking a second medical opinion on his injured leg without New York's permission.[7]

How should we understand these? Can there be facts about the psychological states of groups, and if so, what sorts of facts ground them?

These questions have seen a surge of interest in the last twenty years. (So much so, it may lead one to fear that social theorists are swinging backward to the outdated assumptions of Mill, Durkheim, and the early Watkins.) A great deal of the recent literature on social ontology concentrates on two topics: group intention and judgment aggregation. The first topic inquires into examples like the ones I just listed—*What does it mean for a group or other social entity to intend something?* The problem of judgment aggregation is also a problem about understanding attitudes of groups. People who work on that topic tend to focus on judgments where there is a lack of consensus in a group. *How,* for instance, *do we aggregate the votes of different members of a population into a group vote? How do we turn the judgments of a hiring committee, or those of a group of judges on an appeals court, into a group judgment? What if they are not in perfect agreement on all the subsidiary questions that go into their decision?* So far I have put off these questions. I have wanted to orient us toward a broader set of facts about groups, not just psychological ones. Now we can return to these and confront this recent literature.

---

[4] All these examples are of groups. Group ontology, though, is just a subset of social ontology. Deborah Tollefsen observes that such attributions are equally common for corporations; see Tollefsen 2002.

[5] *The Guardian,* 22 July 2013.

[6] *National Journal,* 22 July 2013.

[7] AP, 25 July 2013.

Psychological facts about groups are the final brick in the argument against "anthropocentrism" in social theory. Even after the preceding chapters, the ontological individualist might seem to have one remaining refuge: *the **important** social facts are psychological facts about groups. And **these** are grounded individualistically*. We can shut this down. Even psychological facts about groups are grounded by a wide range of facts, not just by facts about individuals. And certainly not just by the attitudes of group members.

Approaching Group Intention

A simple view of group intention has been popularized by the "textualist" school of legal interpretation. This school is most vocally represented by Supreme Court justice Antonin Scalia and by Frank Easterbrook, chief judge of the Seventh Circuit US Court of Appeals. Textualism is the view that only the text of a statute is pertinent to its legal interpretation. In particular, it holds that appeal to "legislative intent"—that is, the intentions of the legislature in enacting the statute—should be excised from legal interpretation.

One of the central arguments for textualism is that we could not use legislative intent for interpretation even if we wanted to, since legislatures rarely if ever have intentions in the first place. "Peer inside the heads of legislators," Easterbrook writes,

> and you find a hodgepodge. Some strive to serve the public interest, but they disagree about where that lies. Some strive for re-election, catering to interest groups and contributors. Most do a little of each. And inside some heads you would find only fantasies challenging the disciples of Sigmund Freud. Intent is elusive for a natural person, fictive for a collective body.[8]

Easterbrook does not deny that each legislator has an individual intention, but these intentions are such a hodgepodge that they do not aggregate into one group intention.[9] Even if there were not quite so much hodgepodge, on Easterbrook's view, groups would still lack intentions.[10] It makes sense to

---

[8] Easterbrook 1994, 68. See also Manning 2005.

[9] It is not the only argument for textualism. A closely related argument is epistemological: even if there were legislative intent, we could not know about it, at least not enough to base our legal interpretations on it. And there is a different metaphysical argument for it: namely, that the law is arrived at through compromise, and even if there is a legislative intent, what the law is is what is enacted, not what is yearned for.

[10] Easterbrook 1983.

ascribe an intention to the group only if everyone in a group intends exactly the same thing. In his view, to say "the group intends J" is no more than shorthand for saying, "all the members of the group intend J." Easterbrook gives a simple, individualistic analysis of "the group intends J." It is a matter of all the individuals in the group having that intention.

For years, however, even reductively inclined philosophers have observed that this cannot be right. For a group to intend J, it is neither necessary nor sufficient that all the members of a group intend J.

Margaret Gilbert, for instance, describes this in the context of two people taking a walk together. Suppose Bob is walking to the bank, and Ann is walking to the bank. And suppose they are walking right next to each other. Despite this, they might not be walking *together*. When you're walking together with someone, Gilbert points out, it is not okay for you to just walk off in a different direction, without first saying something to your companion. If you stop to tie your shoe, it is not okay for your companion to walk on at the original pace. *Walking together* involves certain commitments, norms, and accommodations of the walking companions. In contrast, two people *merely having their own intentions* to walk to the bank, and happening to walk side by side, do not make those commitments, are not subject to those norms, and need not make those accommodations. In order to have the group intention to walk to the bank, it is not sufficient to have two individual intentions to walk to the bank.[11]

Michael Bratman has stressed that these individual intentions are not just insufficient, but also unnecessary. If a group of people has the intention to paint a house, for instance, it is not necessary that each person have an individual intention to paint the house. Among the housepainters, no one needs to intend to paint the house him or herself. Rather, they each need to intend something else: that is, that the *group* paint the house. To ground the fact *The group intends to J* does not require that the members have the intention to J. Instead, Bratman argues, it requires a different intention on the part of the members, *that the **group** J*.

Bratman on Group Intention

Building on these observations, Bratman has proposed a theory of group intention, which is one of the most sophisticated and widely accepted theories of a psychological fact about a group.[12] The key to group intention, he argues, is

---

[11] Philosophers also use the terms 'shared intention' and 'collective intention' for this.

[12] Bratman takes his account only to give a set of sufficient conditions, leaving it open for there to be other ways for a group to have a group intention. However, he cannot merely be providing a set of sufficient conditions, since he rejects conditions for not being necessary. (See, for instance, the mafia case in Bratman 1993, 103–4.) It may be better to see his account as proposing one

that the individuals intend to accommodate one another in the group's performance of J. When you and I have the group intention to paint a house, I intend to adjust the way I paint to the way you are painting. I intend to work out a plan with you for dividing responsibility. Maybe I will buy the paint and you will buy the brushes. Maybe I will paint the walls and you will paint the shutters and trim. We do not need to have worked out the details of the plan in order to have the group intention. Nor do we need to have figured out our individual "subplans." Nor do we need to be aware of the details of one another's subplans. But we do need to intend that we, as a pair, will paint the house, and that whatever subplans we work out will "mesh" with the subplans of the other.

Thus Bratman's conditions for **The group intends to J** involves three components:

1. All the people in the group intend that the group J.
2. All the people in the group intend it in a particular way: namely, that the group J by way of meshing subplans.
3. All the people in the group have common knowledge of (1) and (2).

Even without elaborating the notions of "intending *that*," "meshing subplans," and "common knowledge," the essence of his view is reasonably intuitive.[13] Bratman cashes out the notion of a group intention in terms of various intentions of the members, and knowledge about those intentions. There is nothing more to the group intention than the various interlocking individual intentions. Each individual's intention is different from the intention to J, but together they form a unified intention for the group to J.

The Fundamental Role of Group Intention

Bratman conceives of group intention as performing a role for groups similar to the role that individual intention plays for individual people. Individual intention, in Bratman's view, is an attitude an individual has, which plays some role in controlling the person's actions. It is an attitude the individual tends to hold onto, without rethinking it over and over again. And an intention is an attitude that figures into reasoning, deliberation, and planning.

When I form the intention to make a pot of coffee, that intention guides my subsequent actions. The intention might be formed after some process of reasoning about my desires and what would satisfy them. And then that intention

---

determination principle for group intentions, while making no claims about the dependence of group intention on those particular grounds.

[13] See Bratman 1993, 1997. On common knowledge, see Lewis 1969.

guides what I do next—heat the water, warm the Chemex, weigh and grind the beans, and so on.

Group intention, he claims, does similar work. Suppose the members of a group all intend that the group paint the house, all intend that it occur by way of meshing subplans, and all know about those intentions. Then the actions of the group will be guided by those intentions and that knowledge, in the same way that my coffee-making actions were guided by my individual intention. Moreover, that group intention will be stable: it will tend to be held onto. And it will be the basis of reasoning, deliberation, and planning. Bratman points out, for instance, that once a group has that intention, it might bargain about who is to play what role, and about how best to develop the mutual sub-plans. The process of negotiation might be difficult, but still, the group can retain the relevant group intention. (The group might also come to an impasse, and abandon that group intention. But that, too, is analogous to what happens sometimes with our individual intentions, when our planning hits a roadblock. I might abandon my intention to make coffee if I find that the grinder is broken.)

Properties that Both Individuals and Groups Can Have

Let us compare Easterbrook to Bratman. Easterbrook holds that for the group to intend J, all the members must intend J. Bratman's analysis is subtler: he argues that it is neither necessary nor sufficient for the members to intend J. In this sense, Bratman's view is more accommodating than Easterbrook's. Bratman does not require that *any* members intend J, in order for the group to intend it. Instead, he thinks that different intentions are at work, the ones involving meshing subplans.

Still, Bratman's account agrees with Easterbrook's in two respects. One is they both involve unanimity. According to Bratman, the members of the group do not have to individually intend J, but still they do need to be unanimous in a sense. *All* the members must have the intention that the group J (condition 1), and they all have to have it in a certain way (condition 2). The group members also all have to have knowledge of the mutual intentions (condition 3). If legislatures normally have a "hodgepodge" of intentions, as Easterbrook claims, then Bratman must agree with Easterbrook that the legislature rarely has group intentions.[14] This is a fairly obvious place where the defender of group intention might challenge Bratman's view.

---

[14] Or else, Bratman can say that this is a case in which his analysis does not apply.

Easterbrook and Bratman also agree, however, on a less obvious aspect. Although they disagree on *which* individual intentions ground a group intention, Easterbrook and Bratman do agree that group intention is exclusively grounded by the attitudes of the members. And mostly by member intentions, though Bratman also adds "common knowledge" of the intentions into the mix.

The fallacy of decomposition is something you can find in any textbook on critical thinking. This is the fallacy of inferring that because a composite has property P, its parts also have property P. For instance, just because water is liquid, one cannot infer that water molecules are liquid. This is usually an obvious fallacy, but still it can be tempting when we are talking about properties like *intending to J*, which both groups and individuals can have. It seems possible that Easterbrook falls victim to the fallacy, if he believes that for a group to intend J, all the members must have that intention. Bratman, to be sure, does not fall victim to it. He does not "decompose" the property *having intention J* held by a group into the same property held by the members.

And yet, Bratman takes that group property to "decompose" into a very *similar* set of properties. That is, mostly into other intentions, plus a bit of knowledge. He is far from alone in this view. In fact, in considering group attitudes, it is almost universally assumed that they are grounded by individual attitudes. This is so intuitive, to so many people, that it will take three chapters to fully debunk it.

In the rest of this chapter, I "soften the target," by taking some steps to show that this assumption is not as obvious as it may seem. In the next chapter, I turn to group action, showing that the actions of a group are not exhaustively determined by the actions of the group members. Finally, I return to intention, showing that group intentions are not exhaustively determined by member attitudes.

To "soften the target," I consider this general question: Suppose a group has a given property P, which both groups and individuals can have. Do the members need to have it? Do they need to have some property that is closely related to P?

There is no single answer. It all depends on the property. Sometimes, properties work just as Easterbrook claims: for the group to have it, all the members must. Very occasionally, properties of groups work as Bratman's theory claims. But other properties are entirely different. Sometimes, nothing even remotely related to P is required of members in order for the group to have P. Sometimes the group having P does not have anything to do with the members at all. Before making an assumption about how group attitudes *must* work, let us survey the possibilities more generally.[15]

---

[15] All the patterns I will discuss involve the group property and the member properties holding synchronically. It is also worth considering cases in which group properties are grounded by historical properties of the group members and of other objects.

Seven Grounding Patterns

Pattern 1: Perfect Match

Some properties do operate in the way Easterbrook describes intention operating. That is, for a group to have the property depends on the members having exactly that property.

For example, take the property *being completely dressed in red*. Suppose the Red Team has this property. For this to be so, all the members of the Red Team have to have that very property: each member has to be completely dressed in red. Figure 14A depicts the grounds for the fact ***The Red Team is completely dressed in red at time t***:

Figure 14A Example of "perfect match" pattern

A slightly more abstract depiction may be better, to discern the pattern (figure 14B):

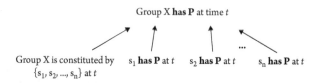

Figure 14B Perfect match pattern

I should stress that even for a fact like ***The Red Team is completely dressed in red at time t***, there are *two* kinds of facts grounding it. One is the "perfect match": that is, that all the people in the set constituting the group have that very same property P at t. Also grounding it—easy to overlook but just as important—is the fact ***The group is constituted by that set at t***. As I have noted, the constitution of a group is often grounded by a heterogeneous list of facts. Among the grounds might be the satisfaction of initiation conditions, maintenance conditions, exit conditions, interleaved facts about the existence of the group over time, and so on. This means that even facts fitting Pattern 1 may depend on a wide range of other facts about the world, because they partially depend on the fact about the set that constitutes them.

Still, once we fix the group's constitution, facts fitting this pattern are grounded very simply. The property holds of the group if and only if it holds of all the members.

Pattern 2: Function of Individual Values

How much does the Red Team weigh? The weight of the group is not the same as the weight of the members of the group. It is the sum of their weights. A fact like **The Red Team weighs 470 lbs. at t**, then, will not fit the "perfect match" pattern. After all, the individual members do not each weigh 470 lbs.

Cases like this fit a different pattern. Instead of having the same property that the group has, the members have properties that "add up" to the property of the group. Properties, like *weighing 470 lbs. at t*, however, are not numbers, so they cannot be added up in a sum. Instead, we treat weight as a *variable* or a *parameter*, that is, as something that takes different numerical values.[16] So instead of talking about the Red Team having the property *weighing 470 lbs. at t*, we can talk about the value of *the weight of* variable applied to the Red Team. For people too, *the weight of* variable has a value. The value of the variable in the grounded fact about the group is calculated just by adding up the values in the corresponding facts about the members.[17] This pattern is depicted in figure 14C:

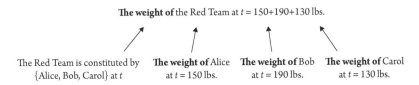

Figure 14C Simple sum pattern

As in the previous pattern, two sorts of facts are involved in grounding. The first are facts about values of variables corresponding to properties of the members. The second is the same fact as above: the fact about what set constitutes the group. Even a simple fact like **The Red Team weighs 470 lbs. at t** depends not just on the weights of the Alice, Bob, and Carol, but also on the fact that the Red Team is constituted by those people.

---

[16] See Strevens 2007, 235, for a more precise way to think of this.
[17] Some variable values can just be added up, to obtain the value for the group. (This also depends on how a given property is parameterized.) In other cases, the value for the group is a more complex function of the member values.

Pattern 3: Recombined Grounds

Other facts exhibit a distinct pattern. Consider, for instance, the fact **The Red Team has a free throw percentage of 67**. The free throw percentage (ftp), in basketball, is the number of free throws made, divided by the number of free throws attempted. Individuals have free throw percentages, and teams do as well. But the free throw percentage for the team is not only grounded by the free throw percentage for the individuals. One person may have only attempted a few free throws, while another person may have attempted a lot. So the team could have a free throw percentage of 67 over some time interval, even if Alice's percentage is 100, Bob's is 100, and Carol's is 66, if Carol attempted 200 free throws, while Alice and Bob attempted only one each.[18]

What grounds the team's free throw percentage is not the free throw percentage of the individuals, but the *grounds* for those percentages. That is, the team percentage is determined by the individual free throw attempts and successes, not the individual percentages. This pattern is depicted in figure 14D:

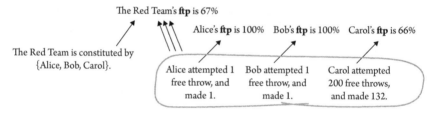

Figure 14D Recombined grounds pattern

In the diagram, the circled facts are among the grounds for the team's free throw percentage. The circled facts are also the grounds for the players' free throw percentages. But the players' free throw percentages themselves do not ground the team's free throw percentage.

There is, of course, a close relation between the free throw percentage of the players and the one for the team. But what grounds the value for the team are not the combined values of that very variable for the players. Instead, is it some *other* facts about the players—in particular, the grounds for those member percentages—that we recombine as a whole, to ground the value for the team.

---

[18] We can use the individual percentages to calculate the team's, by weighting the individual percentages. But that weighting is not grounded by the individuals' percentages alone.

Pattern 4: Grounds of the Same Kind or Same Genus

Bratman's theory does not fit any of the previous patterns. A group's intention to J, he argues, is not grounded by the member intentions to J, nor by the grounds of those intentions. Instead, it is grounded by *different* individual intentions. In other words, facts about the group having one *kind* of property are grounded by facts about the members having *other* properties *of the same kind*. The group intention to J is not grounded by member intentions to J, but it is still grounded by member intentions.

Or at least mostly by intentions. Common knowledge is also involved in Bratman's theory. So he takes group intention to be grounded mostly by various individual intentions, with a bit of knowledge sprinkled in. We might think of Bratman's theory as a blend of two purer patterns:

- a pattern in which group intention is grounded exclusively by *other intentions* of the members. That is, in which a group attitude of kind K is grounded by member attitudes of the very same kind K;
- a pattern in which group intention is grounded by *various attitudes* of the members, without privileging intentions at all. That is, in which a group attitude of kind K is grounded by a variety of member attitudes, which may or may not be of kind K.

Analyses of group intention, judgment, and other attitudes often drift indecisively among these two patterns and their blend. To keep things clear, I will draw figures for all three varieties.

Let K be a kind of attitude, such as intention, belief, judgment, and so on. Then in the first pure pattern, a group attitude of kind K is grounded by member attitudes *of kind K* (figure 14E):

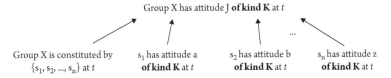

Figure 14E Attitude grounded by member attitudes of the same kind

In figure 14E, the fact that the group has an attitude of kind K is grounded by two sorts of facts: that the group is constituted as it is, and facts about member attitudes only of kind K.

Figure 14F depicts the other purer pattern. Here, a group attitude of kind K is grounded by member attitudes more generally. No particular privilege is given to member attitudes of kind K:

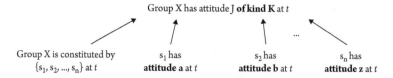

Figure 14F Attitude grounded by various member attitudes

Bratman's theory is partway between figures 14E and 14F. It takes group attitudes of kind K to be grounded mostly by facts about the member attitudes of kind K, with a few other attitudes playing supporting roles. Figure 14G depicts this blended pattern:

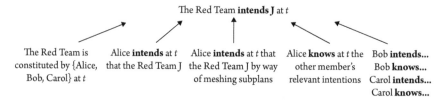

Figure 14G Bratman's theory: grounds of (mostly) the same kind

One question we might ask is whether any *other* properties of groups, apart from group attitudes, conform to patterns like these three. If these patterns are so intuitive for group attitudes, perhaps we should expect that they will work similarly for other properties. Here I can only offer anecdotal evidence: I have struggled to find an example of even one other property that fits a pattern resembling these three variants. For all the other patterns I consider—including the ones to come—examples are easy to find. Not so much for the current patterns. Here is one possibility:

Suppose Alice can lift a 90 lb. dumbbell, Bob can lift a 100 lb. dumbbell, and Carol can lift a 110 lb. dumbbell. The Red Team together, however, cannot lift a 300 lb. dumbbell, since they all have to crowd around it. Instead, the group has the property *can lift a 250 lb. dumbbell*. Moreover, because they have to crowd around it, they use somewhat different muscles and bones than they do when lifting on their own. So their ability as a group to lift a 250 lb. dumbbell is grounded by *other* lifting abilities they have, not exactly their individual dumbbell-lifting abilities. It is still facts about the lifting abilities that ground the team's ability to lift a 250 lb. dumbbell. That is, facts of the same kind or genus. But different facts, not the ones that ground their having the individual ability.

If this description makes sense, the case might fit one of the variants of Pattern 4. **The Red Team can lift a 250 lb. dumbbell** would be a fact about the group, grounded by different facts of the same type about the members. But

I am not even sure this example works: the team's lifting ability might be partially grounded by facts that are not facts about the individuals' lifting abilities. Instead, it might be grounded by facts about various muscles and bones—various grounds of various lifting abilities, rather than by various individual lifting abilities themselves. So instead, this case may fit a pattern more like Pattern 3.

The point, of course, is to wonder if this pattern is as natural as it might seem. Do group intentions and other attitudes fit one of these variants of Pattern 4? I suggest we approach this pattern with caution, and see if there is some good reason that group attitudes should be grounded in this way.

Pattern 5: Intrinsic and Spatial Properties of the Members

At halftime, the high school marching band arranges itself into various shapes. It might, for instance, arrange itself in the shape of the letter 'Y', in the course of spelling out some word as it marches around the field.

A person can also arrange herself into the shape of the letter 'Y' (think of the dance for the song "YMCA"). But of course, for the band to be arranged in the shape of a 'Y', it is irrelevant whether the individual people are. Though the individuals and the group can both have the property *being arranged in the shape of the letter 'Y'*, the two have little to do with one another.

On the other hand, certain facts about the members need to obtain, in order for the band being arranged in that shape. The shape the band is arranged in is a matter of where the members are located. What grounds the fact about the group are intrinsic facts about the members, together with spatial relations among them. But the intrinsic facts that matter have little if anything to do with how the individuals arrange their body parts.

This property fits the following pattern (figure 14H).

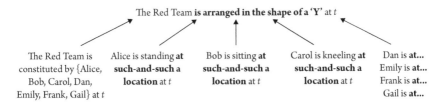

Figure 14H Intrinsic and spatial properties of the members

Being arranged in the shape of a 'Y' is a property that both a group and an individual can have. But for the group to be arranged in that shape, it is irrelevant whether each individual is standing with hands raised. The group having the property is grounded by the members' intrinsic and spatial properties—but not the corresponding property.

So long as I am talking about intrinsic properties of the members of the group, I should highlight the point about constitution again. The property *is arranged in the shape of a 'Y'* is plausibly an intrinsic property.[19] But the fact **The Red Team is arranged in the shape of a 'Y'** is only *partially* grounded by intrinsic and spatial properties of the members. The fact itself is also partially grounded by the fact about constitution, which can be grounded by all kinds of facts. All these other facts can also figure into whether **The Red Team is arranged in the shape of a 'Y'** obtains.

This is tied to the point I made in chapter 5, about the problem with Jaegwon Kim's treatment of facts. Just because a property P is intrinsic does not imply that the fact ***a is P*** is grounded by intrinsic facts about *a*'s constitution. There may be certain facts about groups that *are* fully grounded by intrinsic facts about the group members. But for this to be so, even the constitution and existence of the group need to be grounded in this way. That is, such facts would need to conform to the following pattern (figure 14I):

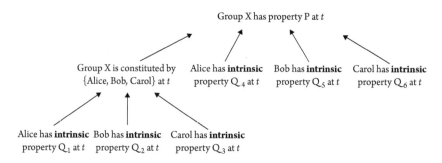

Figure 14I Fully intrinsic fact

In figure 14I, all the grounds are intrinsic facts.[20] The fact about the group's constitution is still one of the grounds for the fact at the top, but that fact is grounded by intrinsic facts about the members. The occasional group might be like this—I return to this point in chapter 17.

---

[19] It may seem strange to say this, since the group having that property depends on the spatial relations among the members (which are extrinsic to them). It still, however, can plausibly be intrinsic to the entity that constitutes the group. This raises two difficult issues: whether it is correct to take a group to be constituted by a set, or rather some other entity; and how to understand 'intrinsic', especially as it applies to objects like groups.

[20] Actually, even these might not be intrinsic, if individuals are not intrinsically individuated. See Kripke (1972) 1980, 113.

Pattern 6: A Mix of Grounds

So far we have worked through permutations of facts about group constitution and facts about its members. But these do not exhaust the grounding patterns. Certain facts about groups can depend on other facts altogether. And some of these need not have anything to do with group members, or with people at all.

We have, of course, seen this several times in recent chapters. We anchor a fact like **The Committee is in session at t** to have more or less whatever grounding conditions we please (figure 14J):

Figure 14J The committee is in session

Being in session, however, is not a property that *both* groups and individual people can have. So it is not quite an instance of what we have been examining with these patterns. Here we are taking a property P that both groups and individuals can have. And considering the fact **Group x has property P**. What sorts of facts can ground such a fact? Can it be grounded heterogeneously?

We are asking this, of course, because we are interested in facts like **Group x performs action A at t**, or **Group x has intention J at t**. Both individuals and groups can perform actions and have intentions. If these facts can have a mix of grounds, one way they might come out is depicted in figure 14K:

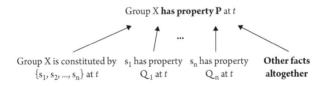

Figure 14K A mix of grounds

Notice that in the figure, we do not assume that the members have the very same property that the group does. As we have seen, just because property P is one that both groups and individuals can have, that does not mean that the group having P depends on the members having it.

Do group actions and attitudes conform to this pattern? This is the topic of the next two chapters.

Pattern 7: The Members Are Irrelevant

In all the patterns so far, the members play at least *some* role in grounding the property of the group. Some properties of groups, however, are not grounded by anything about the members whatsoever.

Consider, for instance, *age*. To figure out something's age, you figure out the time since it came into existence. Now consider the age of an intramural basketball team, like the ones we discussed in chapter 13. The Red Team, for instance, came into existence two months ago, when the manager and approver took the initiating actions. But that has nothing to do with the ages of the members: each of the members of the Red Team came into existence around nineteen years ago. In grounding the age of the group, the age of the members is irrelevant. Nor is anything else about the members relevant.

If you want to know the age of an individual person, you would just look at the person's own history. Why do facts about the manager figure into facts about the age of the team, whereas facts about the age of an individual person are grounded by the person's internal history? It is simply because the age of something depends on when it came to exist. The existence of an intramural basketball team depends on its having been initiated by a manager in the appropriate way. A person's age also depends on when the person came to exist, but a person's existence does not ontologically depend on external factors in the same way (figure 14L).[21]

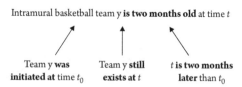

Figure 14L Age of a team

---

[21] Again, persons may be extrinsically individuated (see Kripke [1972] 1980, 113, and the succeeding literature on the essentiality of origins). Still, there is a closer tie between the existence of a person and her intrinsic properties than there is between the existence of a group and its intrinsic properties.

That does not mean that a property like *being two months old* is applied in one way to individuals and in a different way to intramural basketball teams. How we apply *being two months old* to an object—any object—is to look at the things in the world that ground its coming into existence. This is the same for an individual and for a group.

This example has an important moral: it can be deceptive to consider only a limited set of cases. If we were only to look at age as it applies to individuals, we might mistakenly infer that it is intrinsic. And if we were only to look at age as it applies to groups that form at the birth of the members, we might mistakenly infer that the age of the group ontologically depends on the ages of the members. But those are special cases.

Being two months old is not the only property for which the members are irrelevant to the group's having it. Other properties are like this too—historical properties of a group, certain moral properties, properties having to do with powers and jurisdictions. The idea that any property of a group must be grounded by related properties of the members just reflects a failure of imagination.

It is too easy to assume a close relation between a group possessing a property P and its members possessing P (or something related to P). My aim in this chapter has been to soften these assumptions. In the next chapter, I turn to a more direct treatment of group action and intention. But let me conclude this chapter by pointing out some flaws in a prominent recent discussion of the relation between group and individual attitudes.

List and Pettit on Group Attitudes

It is commonly taken as obvious that group attitudes fit one of the variants of Pattern 4. In a prominent recent book, *Group Agency*, Christian List and Philip Pettit give an explicit argument for it.[22]

The Starting Assumption

In a section titled "The Supervenience of a Group Agent," List and Pettit argue that group attitudes supervene on the attitudes of individuals. They begin with the following:

---

[22] Raimo Tuomela gives a different argument for the same point in Tuomela 1995, 256–9. I discuss some issues with Tuomela's assumptions in chapters 17 and 18.

> The things a group agent does are clearly determined by the things its members do; they cannot emerge independently. In particular, no group agent can form intentional attitudes without these being determined, in one way or other, by certain contributions of its members, and no group agent can act without one or more of its members acting.[23]

To clarify, when List and Pettit say "group intentional attitudes," they mean the same thing as "group attitudes." That is, not just intentions, but also beliefs, judgments, knowledge, and so on.

This passage gives their starting assumption: group attitudes are determined by "contributions" of the group members. It is not entirely clear what this means. A natural interpretation is that group attitudes conform to at least Pattern 5: they are exhaustively determined by intrinsic and spatial properties of group members. On the other hand, Pettit elsewhere has argued for an "externalist" view of attitudes, that is, that attitudes of individual people are partly determined by facts about the external world.[24] So presumably the contributions of group members can include these "wide" attitudes.

Still, we do not want to interpret List and Pettit as *assuming* that group attitudes are exhaustively determined by "wide" member attitudes: this, after all, is what they aim to show. So to accommodate this point, we should take their starting point to be this: group attitudes fit an expanded version of Pattern 5. That is, expanding the intrinsic and spatial properties of members to include "wide" attitudes as well.

The Argument

The core of List and Pettit's argument is this: it is logically possible that group attitudes are determined by something other than individual attitudes. But this possibility is "unrealistic."

To argue this, they discuss a thought experiment described by Ned Block in the late 1970s, the "China-body system." In this thought experiment, each of the billion people in China is imagined to follow a set of mechanical procedures in which each person replicates the activities of a handful of neurons from a real person's brain. All the people in the country set up mechanical interconnections with one another, to match how the corresponding neurons would interact with one another. The whole populace, acting together, replicates the neural activity of a human brain. Block points out that this system

---

[23] List and Pettit 2011, 64. They also make this argument in List and Pettit 2006.
[24] Pettit 1993.

can be understood as performing the same "functional roles" as a human brain does. For a given set of stimuli or inputs, the "China-body" will produce the same outputs as a brain does.

List and Pettit point out that the group attitudes of the "China-body" are not determined by the attitudes of the individual Chinese players. Rather, they are determined by their mechanical interconnections. "The attitudes of the group agent," comment List and Pettit, "clearly supervene on the contributions of individuals, but only on their non-attitudinal contributions."

"Nevertheless," they go on,

> this possibility is not a very realistic one. In most real-world examples of jointly intentional group agents, the members' actions and dispositions sustain the group's organizational structure, but under that structure the group's intentional attitudes are formed on the basis of the intentional attitudes that members manifest.[25]

Diagnosing the Argument

This strategy is puzzling. For one thing, Block himself never claimed that the China-body has attitudes—in fact, the point of the example is to deny it. The target of Block's article is a "functionalist" theory of the mind. According to Block, such a theory would wrongly insist that the China-body has attitudes. It would "classify systems that lack mentality as having mentality."[26] Given this, it is not obvious why List and Pettit even consider the example.

This might seem to buttress their argument: List and Pettit are surely right that the China-body is unrealistic, and on top of that, it may not even have attitudes anyway. Yet even if we grant both of these, it is still not clear that List and Pettit can draw the conclusion they do. The steps of the argument as a whole seem to be these: (1) Group attitudes clearly conform to an expanded Pattern 5. (2) Here is one bizarre way a group attitude might conform to the expanded Pattern 5 without conforming to Pattern 4. (3) But that way is bizarre. (4) So, group attitudes conform to Pattern 4.

But that isn't an argument. So what do List and Pettit have in mind? I think what may be going on is that they take two different alternatives to exhaust the possibilities. Either a group is made up of individuals playing a role as intentional, agential, thinking members, in which case group attitudes are determined by individual attitudes. Or else a group is made up of individuals

---

[25] List and Pettit 2011, 65.
[26] Block 1980, 275.

playing a non-intentional, non-agential, non-thinking role. In that case, the group looks something like the "China-body."[27] Since people clearly do act as thinking agents, even as group members, only the former possibility is realistic.

These alternatives, however, are too stark. As we have seen, a group is constituted by individual people, yet facts about those people typically play only a partial role in determining facts about the group. This suggests that even if member attitudes do *partially* ground group attitudes, member attitudes may not *fully* ground group attitudes. The role members play need not be exclusively attitudinal or exclusively mechanical: members may play both sorts of roles. And on top of that, there might be other facts that also figure into the grounds of group attitudes.

When it comes to List and Pettit's actual theorizing about group judgment, the assumptions get still stronger. The argument above—flawed as it is—is an argument that group attitudes conform to the broadest version of Pattern 4. That is, it argues that group intentions, judgments, beliefs, and so on, are grounded by *some variety* of attitudes of individuals. However, when it comes to filling out the theory, they assume more. They do not just assume the supervenience of group judgment on individual attitudes. Nor do they just assume that group judgments supervene on *various* individual judgments. Instead, they assume that group judgments about a set of propositions are determined by individual judgments about *that* set of propositions, plus a collective decision procedure.[28] In other words, when they get down to business, they take group judgment to fit an even more restrictive pattern, closer to Pattern 2.

Unlike many others, List and Pettit argue their supervenience claim, rather than just assuming it. But each of the steps is worrisome. It is not clear that we can assume from the outset that group attitudes conform to an expanded Pattern 5 (or however else we should interpret the assumption that group attitudes are "clearly" determined by the contributions of the members). The argument from Pattern 5 to Pattern 4 does not work, unless we assume that only two stark alternatives are available. And it is not clear why we should move from the broad version of Pattern 4 to a variant on Pattern 2. Of course, even if their argument is not bulletproof, that does not mean that their claims are mistaken. So let us turn to more direct scrutiny of group actions and intentions.

---

[27] Along the same lines, they also use the example of swarms of insects in List and Pettit 2006.

[28] List and Pettit 2006, 2011. They emphasize that group judgments supervene "set-wise" rather than "proposition-wise": that is, that they depend on aggregation procedures applied to sets of individual judgments of members, and that we cannot expect that a particular group judgment will supervene on the corresponding individual judgment. Still, the supervenience is not on just any individual judgments: it is on individual judgments about the set in question.

15

Group Action: More than Member Action

What more can there be to group action, apart from the actions of the members? What more to group intention than the attitudes of the members? These questions almost seem rhetorical. After all, you can either build groups out of their members, or else you can believe in crazy social spirits. Maybe there are other unrealistic alternatives, like the "China-body," but these are hardly worth considering.

Although this reasoning is deeply held, it is mistaken. After our work in earlier chapters, we can see why: facts about group actions and intentions are anchored to have more heterogeneous grounding conditions. Facts about group members are just part of the grounds.[1] The aim of this chapter and the next is to argue this point directly.

I will begin with group action, and discuss three different categories of nonmember grounds. Each category, on its own, proves the point. But taking all three together, we see that these nonmember grounds are not accidents or quirks we can overlook. Instead, they are built deeply into the design of groups, to powerful effect. Over thousands of years of sociality, we have figured out—consciously or not—how to craft groups to accomplish our desired ends, sometimes despite rather than because of their members.

Action and Intention are Intertwined

We start with action not because action is simpler or more fundamental than intention. Action and intention are intertwined. When I form an intention to

---

[1] When I say "individual action" and "group action," I am referring to the intentional actions of individuals and groups, as opposed to reflexes, behaviors, happenstance byproducts, and other things that are not full-blooded actions.

brew a pot of coffee, for instance, that intention does a job for me. The intention alone does not give me my caffeine fix, but it is a step along the way. The intention is part of my system of deliberating, planning, and acting. In forming that intention, I commit myself to the task. I can revoke my commitment, should I decide to. But usually it guides the next steps I take, as well as the decisions I make along the way.

My individual intention is just a part of my overall system of practical activity. It connects with other attitudes of mine, such as my beliefs and desires. It also connects with my reasoning faculties, with the actions I take over time, and with the appliances in my kitchen. All these gears move together in my practical coffee-making activity.

Actions and intentions, in other words, are interlocking components of a system of practical activity, a system that also includes reasoning, planning, responsibility, and more. A full account of action will involve all these components, including intentions, just as an account of intention will involve action. All the roles involved in the system as a whole—the roles of actions, intentions, plans, and so on—figure into anchoring the parts so that they play well with the others.

Individual intention and action play these roles in an individual person's system of practical activity. Group intention and action play much the same roles for groups. If we can make sense of group intention and action at all, it is in the context of a similar system—a network of group attitudes, group reasoning and planning, and group action. Group intention and action are not likely to parallel individual intention and action in every way: after all, groups and individuals are very different things. Still, any theory of group intention and action needs to make sense of them doing roughly analogous jobs.

So why start with action? In certain ways, action is less opaque than intention. Actions are easier to observe. And when it comes to groups in particular, actions have another advantage: in many cases, we *explicitly* anchor some of the grounding conditions for group action. That is, we explicitly set up conditions under which a group can act. This gives us a clear window into at least some of the grounds for group action.

We cannot just anchor group action however we want: it is intertwined with group intention, and with other parts of our system of practical activity. But that does not mean we are powerless to shape the grounding conditions for group action. We often anchor direct and indirect constraints on it. In this chapter, we will examine three of these ways.

# 1. Unequal Contributions to Group Action

A common way to mold the grounding conditions for group action is to set up hierarchies, divisions of labor, and structures of power. That is, to anchor members to make unequal contributions to the group's actions.

This is an obvious feature of groups. It is not as obvious, however, why it leads to the failure of member action fully determining group action. Even if group members make unequal contributions, it is still group members who fill the roles and have the powers. So it might seem that unequal member contributions cannot imply that additional factors are involved in determining group action. But they can.

To illustrate the point clearly, I will give four brief examples. Basically, all we need to do is draw grounding patterns, as in the last chapter. Still, we have to be careful. In the first two examples (A1 and A2), unequal contributions to group action do *not* imply that group action is determined by more than member action. The first two cases, that is, *are compatible* with member action exhaustively determining group action.

In the second two cases (B1 and B2) unequal contributions *do* imply that group action depends on more. In these two cases, group action is not exhaustively determined by member action.

A1: The Student Government

Suppose that Mrs. O'Leary authorizes her 9th grade class to have a student government, and writes the rules for it. In enacting those rules, she anchors the conditions for *x is class president*. To be class president, a member of the class must win an election, voted on by the students in the class. Mrs. O'Leary also anchors the president to have disproportionate voting power on decisions made by the class. The president gets five votes on every decision, while the other members of the class get only one vote each.

The class conducts its elections, and elects Jill as class president. Subsequently, the class decides on issue J. The anchoring and grounding of the fact are depicted in figure 15A. In the figure, I have highlighted in bold the facts that depend on more than just the members of the class.

The anchors on the left are actions taken by Mrs. O'Leary, not the actions of students. But I only mention them to point out their irrelevance to the present question. We are considering whether the fact **The class votes yes on J** is partly *grounded* by facts about nonmembers. As I have argued,

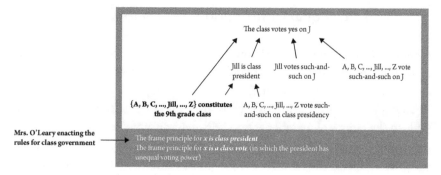

Figure 15A The student government

the anchors of a frame principle are not among its grounding conditions. Therefore, even though this frame principle is anchored by Mrs. O'Leary, that does not show that facts about Mrs. O'Leary are among the grounds of **The class votes yes on J**. (Notice that for the conjunctivist, the determination of group action by member action fails even here. The conjunctivist includes the anchors among the grounds.[2] So the conjunctivist will hold that in this case, certain facts about Mrs. O'Leary also figure into grounding the class vote.)

Instead, the fact **The class votes yes on J** is grounded by four facts, as depicted inside the frame: the fact about who the members of the class are, the fact Jill is class president, Jill's vote, and the votes of the other members of the class. (Remember that the disproportionate weight of the president's vote is anchored in the frame.)

The only fact in bold, in the frame, is **{A, B, C, ..., Jill, ..., Z} constitutes the 9th grade class**. This fact, of course, is grounded by more than the students themselves. It is grounded by whatever facts are involved in school enrollment. But once we are given that fact—i.e., once we fix class membership—the remaining grounds are nothing more than facts about the class members. In particular, the fact **Jill is class president** is grounded by actions of the members of the class themselves (i.e., their earlier votes). Altogether, the group vote is grounded by member actions.[3]

---

[2] See chapter 9.

[3] If Jill's being president depends on member attitudes as well as votes, then strictly speaking group action depends on more than member action altogether: it also depends on member attitudes.

A2: The US House of Representatives

Let's reinforce this with a case that shows the exact same thing. It is another group with a hierarchy, where group action is plausibly determined by member action alone.

The Speaker of the House of Representatives has disproportionate power in the House. When the House performs an action, the actions of the Speaker contribute differentially to the actions of the House. And as in the previous case, it is the members themselves who elect the Speaker.

Figure 15B depicts the grounds for an action of the House, taking into account the differential contributions of the Speaker. This example has the same structure as the previous one. (In this diagram, I am just including what takes place inside the frame. Since the anchors are not among the grounds, they are irrelevant.)

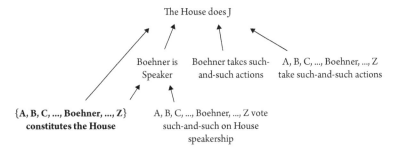

Figure 15B The House of Representatives

As in the previous case, facts about nonmembers figure into grounding membership in the House. But here again, once we fix the membership of the House, it is those members who select the Speaker. So altogether, the group action of the House is fully grounded by actions of the members.

B1: Microsoft Stockholders

Now to the other cases. All it takes is one more ingredient for supervenience to fail. In cases A1 and A2, the group members choose the class president and House Speaker. But if those choices are *not* made by the members alone, then group action fails to depend on member action.

Consider the Microsoft stockholders. I own 100 shares of Microsoft stock, and Bill Gates owns 420,000,000. When he and I vote at the annual stockholders meeting, his vote carries 4,200,000 times the weight of mine. This is how we typically anchor voting rights for corporations: voting power is in proportion to stock ownership. Suppose that all the stockholders of Microsoft, apart from Bill and me, are split evenly on some question. I vote one way, and Bill votes the other. In that case, Bill's side wins the vote. (Some corporations operate differently. Mark Zuckerberg owns 18 percent of Facebook, but controls 57 percent of its voting rights. This is simply the result of anchoring group votes to have different grounding conditions.)

There is a key difference between this and the previous two cases. Consider the facts **I own 100 shares of Microsoft** and **Bill Gates owns 420,000,000 shares**. These facts are not grounded by facts about Bill and me alone, nor even by all the other stockholders. Rather, they are grounded by facts about historical contracts, stockholder agreements, money transfers, stock markets, and so on. In contrast, the fact **Boehner is Speaker of the House** is grounded only by facts about members.

It is not just the differential voting power that defeats the supervenience of group action on member action, in this case. Rather, it is the *assignment* of that differential power to particular people. The fact **The Microsoft stockholder group does J** does not just depend on the fact that someone has differential power. The action depends on the particular assignment, as depicted in figure 15C:

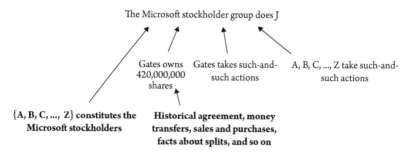

Figure 15C Microsoft stockholders

Suppose we had all the individual actions of Bill Gates and of me. Also, suppose we had the fact that we are both members of the group, that is, both stockholders. And suppose it is anchored that voting power is in proportion to stock ownership. All that is still not enough to determine that Bill's vote carries more weight than mine. To all these, we need to add the grounds for the fact that he owns 420,000,000 shares, and that I own 100. In other words, it is externally grounded *which* members make *which* contributions.

B2: The Supreme Court

One more example to illustrate the point. Supreme Court justices make unequal contributions to some of the court's actions. Among the powers anchored for the chief justice is the power to assign authors for opinions of the court. Suppose Justice Scalia gets too big for his britches, and sends out an email to his colleagues assigning responsibility for writing a case's opinion to Justice Thomas. Suppose that Chief Justice Roberts sends out an email assigning responsibility for the same case to Justice Kagan. Then suppose both Kagan and Thomas publish opinions. Kagan's publication is the issuance of an opinion by the court. Thomas's publication is not.

It is not enough to fix the membership of the Supreme Court and what those members did, in order to determine the action of the court. Even granted this, we need to ground the further fact that Roberts is chief justice (Figure 15D).

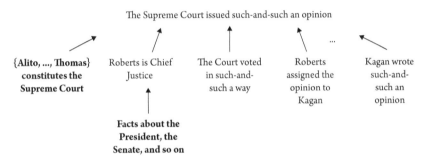

Figure 15D Opinion of the Supreme Court

External facts—not just facts about members—ground the fact that Roberts has differential power. The role of chief justice is anchored by the constitution and Judiciary Acts. But the fact that Roberts fills that role has external grounds. These external grounds are among the facts that ground the issuance of opinions by the Supreme Court.

The difference between A1/A2 and B1/B2 is small. It may seem like a technicality that the actions of a group can depend on more than the actions of members. But it is more than this. It illuminates a powerful and typical way we constrain group action.

It is the norm, not the exception, for different members of a group to be assigned different powers. It is also the norm, not the exception, for these assignments to be grounded externally. Most positions in political hierarchies are grounded by facts about people other than the members. As are positions on sports teams. As are positions in business groups.

Why do this? Why not select the people to be government workers, and then let facts about them ground their place in the hierarchy? Why not give generic contracts to players in the NFL, and let facts about them ground their positions on the team? Why not let the employees in a business team appoint their own leaders or managers?

Sometimes, such assignments *are* grounded by the members. This is what cases A1 and A2 illustrate. But much of the time, that would be a disaster. We often want group action to be sensitive to facts about the external world, and to serve external aims. We want sports teams to win. We want business teams to pursue the aims of upper management. We want stockholder votes to favor the majority owners. To accomplish this, we take some of the power for grounding group action out of the hands of group members.

And notice that this is a point about the *metaphysics* of group action, not just about its causes. These examples do not just show that nonmember facts are involved in *causing* members to be assigned their places in a hierarchy. Rather, they show that nonmember facts are *partial grounds* for those assignments. And therefore, are partial grounds for group action.

2. Direct Constraints on Group Action

For the next category of cases, I turn to a more traditional argument style. (Traditional, at least, among philosophers who think about interlevel metaphysics.) Here I will use the standard argument for supervenience failure. The form of this argument is to describe two near-identical scenarios. The scenarios are carefully described, so that the member actions are the same in both, but the group actions differ from one scenario to the other. This shows that the member actions do not exhaustively determine the group actions.[4]

I also turn to a different kind of constraint we anchor on group action. In the last section, I considered unequal contributions by different members. Those have an *indirect* effect on group action: we do not constrain group action directly, but do so via constraints on the members. In this section, I consider more direct constraints on group action. At the beginning of our inquiry into groups, we examined facts like ***x is in session at time t***, ***x has such-and-such a jurisdiction at t***, and ***x has such-and-such powers at t***. We anchor frame principles for facts like these precisely because these facts figure into the grounds of group action.

As we pointed out, the grounding conditions for facts about activation, jurisdiction, and powers can be wild and heterogeneous. These same heterogeneous grounds, then, also can partially ground facts about group action. It will be helpful to go through at least one traditional nonsupervenience argument in detail.

---

[4] For discussion of this argument form, see McLaughlin 1984. A claim about dependence is not exactly the same as a supervenience claim. But if supervenience fails, then a fortiori, exhaustive dependence fails (see chapter 8).

The Setup

Consider a variation on the MassDOT/MBTA case from chapter 10. Suppose the governor of Massachusetts is setting up these boards for the first time, anchoring various frame principles for them. He anchors powers for the MassDOT board, such as the power to approve toll increases on state highways. And he anchors powers for the MBTA board, such as the power to approve fare increases on the subway.

The governor also anchors conditions for the boards to be in session. Suppose, for instance, he anchors the office of a "parliamentarian" for each of the two boards. The parliamentarian of the MassDOT board is not a member of that board, but is responsible for convening and terminating MassDOT board meetings. And the parliamentarian of the MBTA board is likewise not a member of that board, but is responsible for convening and terminating MBTA board meetings. Finally, suppose that people are appointed to be the members of the boards: Alice, Bob, and Carol are named to both the MassDOT and MBTA boards. And each board is also assigned a parliamentarian: Dorothy and Bertha, respectively.

When Dorothy convenes the MassDOT board meeting, that group has the power to make decisions applying to highways. And when Bertha convenes the MBTA board meeting, that different group with the same members has the power to make decisions about the subway. Conversely, the MBTA board lacks the power to make decisions about highways, and the MassDOT board lacks the power to make decisions about subways.

Two Scenarios

Now let us construct the two scenarios, to show that even when the member actions are indiscernible, the group actions may differ (figure 15E). The first scenario consists of the following sequence of events. Alice, Bob, Carol, Dorothy, and Bertha gather in a room at 10 a.m. Dorothy calls a meeting of the MassDOT board to order, and the members voice their approval of increased highway tolls, following which Dorothy closes the meeting at 11 a.m. Then Bertha calls a meeting of the MBTA board to order, and the members voice their approval of increased subway fares, following which Bertha closes the meeting at noon.

The second scenario is exactly the same, except for one detail: unbeknownst to the members, Dorothy and Bertha convene their respective meetings at the wrong times, so the committee actions therefore misfire. The sequence of events in the second scenario is this: Bertha convenes the MBTA board meeting, and the members voice their approval of increased highway tolls, and then

Bertha closes the meeting. Subsequently Dorothy convenes the MassDOT board meeting, and the members voice their approval of increased subway fares, and then Dorothy closes the meeting.

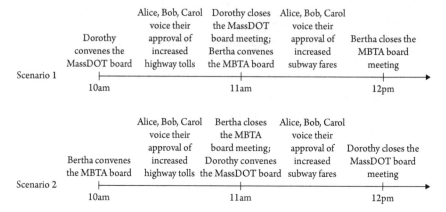

Figure 15E Two scenarios: actions taken and action misfiring

In the second scenario, the committees were convened incorrectly. Therefore, neither committee had the power to take the actions it tried to take. The MassDOT board cannot raise subway fares, nor can the MBTA board raise highway tolls. Thus these attempted actions by the committees, which were straightforward intentional actions in the first scenario, are failed attempts at action in the second scenario.

This is obviously an artificial case, but it illustrates how group action can depend on nonmembers. Alice, Bob, and Carol remain unchanged from scenario 1 to scenario 2. But the group actions are different: they take place successfully in scenario 1, and misfire in scenario 2.

What Is Happening

In each of the scenarios, the MassDOT board and the MBTA board have the same members, and are in the same contexts. The two groups coincide through the entire scenario, including at 10 a.m. and at 11 a.m. and at 12 p.m. Still, it is no mystery what is going on. The actions of the MassDOT board have different grounding conditions than do the actions of the MBTA board. In the first scenario, the grounding conditions for activating the two committees are met. In the second scenario, they are not. And the reason they have different grounding conditions is that the Governor explicitly anchored them that way. He anchored the MassDOT board to be able to act under certain conditions, and the MBTA board to be able to act under others.[5] Figure 15F depicts the grounds for an action of the MassDOT board, with nonmember grounds in bold:

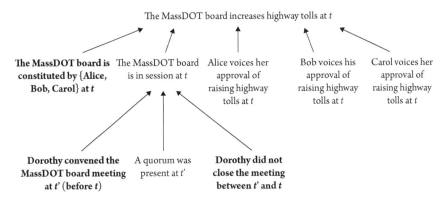

Figure 15F Action of the MassDOT board

We routinely anchor constraints like these. Ordinary committees, ordinary courts, and ordinary boards of directors are typically anchored to have limited spheres of action. It is evident why: we set them up so that they can perform certain roles in certain spheres. Outside of those spheres, we anchor them to be impotent.

The Supreme Court, for instance, is the ultimate arbiter of questions that affect hundreds of millions of people. This would be dangerous if it were not also limited. Since we anchor the court to have these powers, we also tightly circumscribe the actions it can take. Accompanying the powers we anchor for the court, we also anchor constraints on its actions. These constraints take the form of extra grounding conditions for facts of the form *x* **is an action of the Supreme Court**, that involve facts apart from the actions of members.

Notice that we do not have to decide between the members either having iron-fisted control over group action, or else being thoughtless neuron-like cogs, like parts of Block's "China-body." Alice, Bob, and Carol are rational actors, and their actions contribute to the actions of the two committees. It is just that there is more to the story.

---

[5] Karen Bennett points out the presence of this sort of "grounding problem" in Bennett 2004b. One advantage of the grounding–anchoring model is that it addresses this problem.

Four Potential Worries

It may be helpful to address some worries.

1. First is the worry that the argument is right, but irrelevant. The worry is that constraints like these are merely a matter of the "institutional context" a group finds itself in, and are not relevant to determining the actions of the group itself. Facts about nonmembers, it might be thought, are "exogenous" facts that are not pertinent to the dynamics of group action.

This misses the point. Nonmember constraints on action can be just as dynamic as the actions of group members. Recall that the MassDOT board exactly coincides with the MBTA board. If we consider only what the group members do, we will not just make subtle mistakes about the group actions in odd contexts. Rather, we will entirely miss the differences between the actions of the two groups. This point applies to modeling as well. If we want to model the two boards, the actions of Dorothy and Bertha are at least as pertinent to group action as those of Alice, Bob, and Carol.

These external factors *have* to make a practical difference. If constraints like these had no practical effect on group action, we would not bother to anchor them, as we so commonly do. We do a lot of work to circumscribe the actions of groups with external grounds. That would be inexplicable, if all that grounded group action was the members themselves.

2. A second worry is whether Dorothy and Bertha really should not be considered members of the committees. It might seem odd to exclude them, since they play such important roles.

This, however, is not unusual at all. For instance, the Sergeant-at-Arms of the Senate is not a member of the Senate, nor is the minister a member of the couple she marries.

In any case, if this does seem like a problem we can easily rewrite the example to defuse it. As I described it, the committees are convened and dissolved by Dorothy and Bertha. But instead, let them be convened and dissolved by external facts about the world. For instance, suppose that the governor anchors that the MassDOT board is in session for an hour starting at 10 a.m. on Monday, Wednesday, and Friday, and that the MBTA board is in session for an hour starting at 10 a.m. on Tuesday, Thursday, and Saturday. Written that way, it would be the external time of day that would determine which board was in session.

3. A third potential worry is that these phenomena arise from the groups being "social organizations" established by the governor, the legislature, and popular elections. Maybe the determination failure is a product of these external anchors.

But this too is easily addressed. The charters of the committees may be written and approved by Alice, Bob, and Carol themselves, and the actions taken by the respective committees may be limited to ones that they themselves have the power to authorize. The members themselves may anchor the grounds for committee action to be, in part, out of their own hands.

4. A final worry is about the "supervenience base." The form of the argument is to set up two situations and contrast them with one another. Both of the situations are meant to be indiscernible in terms of the member actions. The worry, however, is that in the two situations, the member actions might not actually be indiscernible. It is true that Bob voices his approval indiscernibly in both situations. But another of Bob's actions in the first situation is to *vote* to increase highway tolls. In the second situation, however, Bob did not actually vote: he *tried* to vote, but his individual voting action misfired, just as the group action did. This suggests that the case might not successfully show that the group actions are determined by more than the member actions.[6]

To address this, we need to be clear on just what is supposed to be determining what. When a theorist says that group actions are exhaustively determined by member actions, it is important not to smuggle the group actions back down into the member actions. It is true that Bob does not succeed at voting in scenario 2. But the action *successfully voting to raise highway tolls* is not reasonably included the supervenience base, because it depends on Dorothy's activation of the group. Someone can always make a supervenience claim true by expanding the supervenience base to include the things that are argued to supervene on it. But that just trivializes the claim, and even the defender of supervenience does not want that.

3. Political Systems and Membership Constraints

In preceding chapters, I repeatedly stressed that group *membership* depends on much more than facts about the members. So far I have kept that point to the side. The previous cases show that even granted the membership, the actions of the members may not be enough to determine the actions of the group.

At this point, I want to reopen the grounds of membership. Membership conditions are often the strongest lever we have for shaping group action. In fact, our contemporary political systems are designed on this insight. As I will discuss, group action can be so strongly constrained that the group members become almost irrelevant in determining that action. Instead, conditions on membership can dominate.

---

[6] I am grateful to Sarah Paul for pressing this point.

To describe this point, I will start with an important model in political science: the "electoral control" model.[7] This model is helpful to see how membership conditions can be used to control action. However, most versions of this model have a critical blind spot: they consider only the *causal* effects of the electorate on the membership of a legislature. They do not address an even more powerful mechanism for electoral control: a *constitutive* mechanism. Once we highlight this mechanism, we can see that external membership constraints can have a profound effect on group action over time. And that similar constraints are central to the design of many types of groups, not just political ones.

Electoral Control

A key concern in the design of political systems is to ensure that politicians, such as legislators, act in the interests of their constituents. A legislature may not always do what its constituents want. Politicians are people. Like the rest of us, they have interests. If a politician's interests diverge from the interests of her constituents, then she will need to decide whose interests get short shrift.

People often wonder why their legislators act so badly, but from the perspective of the political theorist, the question is the opposite. Why does the legislature *ever* do what is in the constituents' interests? It just seems irrational for a legislator to act against her own interests. How does an electorate manage to control the actions of the legislature at all?

This is known in economics as a "principal-agent" problem. A principal (the electorate) hires an agent (the legislature) to perform some task that serves the principal's interests. How does the principal get the agent to act in alignment with the principal?

An obvious way for the electorate to get the legislature to do its bidding is to set up a system of rewards or punishments.[8] If the legislature does what the electorate wants, the legislators get a bonus. For legislators, however, this kind of system is difficult to implement. The big problem with direct incentives, monetary ones in particular, is that they tend to stack the outcomes in alignment with the interests of the wealthy, not with those of the electorate as a whole.[9]

---

[7] The predominant models in the field are descendants of those developed by Robert Barro and by John Ferejohn. See Barro 1973; Ferejohn 1986.

[8] Becker and Stigler 1974.

[9] See, for instance, Rose-Ackerman 1978.

Fortunately, in electoral systems there is a different lever for controlling the behavior of legislators. If the electorate is unhappy with the way the legislature is voting, it throws the legislators out of office. In a political system with frequent elections, the replacement of misbehaving elected officials can be powerful. Over time, it can ensure the conformance of the vote of the legislature with the preferences of the population.

How do elections manage to affect the votes of the legislature? The most prominent models focus on one mechanism: the legislators are kept in line by the fear that they will be tossed out. That fear influences their incentives, and hence their decisions. Given that legislators are rational actors, they can be counted on to make choices in their own best interests. When they are in office, legislators receive nice salaries and lots of prestige. To keep their perks, they want to survive the next election cycle. But if they do not do the bidding of the electorate, they will not. So they have an incentive to do what the electorate wants. Figure 15G sketches these causal influences on the votes of the legislature:

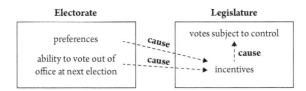

Figure 15G Causal mechanism

A Second Mechanism

There is another mechanism for electoral control that is frequently more powerful than incentive effects.[10] Even if legislators completely resist the incentives that elections generate, the electorate can still control legislative votes *over time*. Suppose we have a set of legislators with very strong convictions, so that they are completely impervious to the preferences of the electorate. They are so stubborn that we cannot change their minds, no matter how much we change their incentives. In that situation, the electorate may be powerless to sway the votes of the legislature closer to their preferences. Until, that is, the next election cycle arrives. At the next election cycle, the electorate regains control: it can choose a different set of people to constitute the legislature.

This, of course, is how electoral systems are mostly designed to work. We do not elect officials expecting them to change as their incentives do. We elect

---

[10] This mechanism is often overlooked, but appears in a few models. See Besley 2006; Besley and Smart 2002; Fearon 1999.

them because we like particular convictions they have, and assume that those convictions are rather fixed. When the electorate no longer shares the convictions of those officials, it swaps them out.[11] Figure 15H sketches this mechanism. In the figure, causal relations are dotted arrows and grounding relations are solid arrows.[12]

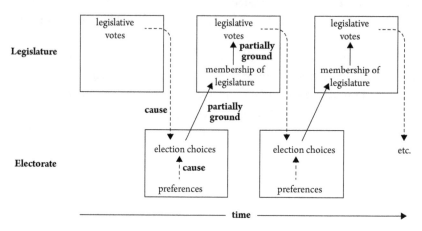

Figure 15H Constitutive mechanism

In the diagram, the choices of the electorate partially ground the membership of the legislature. Which, in turn, partially grounds the votes of the legislature. Causes are also involved in this mechanism: the voter choices are causally influenced by their own preferences. And their choices are informed by their observation of how the legislature voted earlier.[13]

These are two distinct mechanisms for electoral control. The first is a *causal* mechanism: the fears and other expectations of the legislators about the next election have a causal effect on their choices. The second is a *constitutive* mechanism: the electorate changes the constitution of the legislature over time. With the second mechanism, the electorate has the power to mold the actions of the legislature in its image, even if incentives have no effect at all on legislators. It even works if the electorate has *no causal connection* at all

---

[11] Some theorists argue that this second mechanism will be swamped, in the long run, by "office-motivated" politicians. See for instance Calvert 1986. However, see Besley 2006, section 4.4.2, for an argument that even a minimal presence of "types" of politicians undermines the first mechanism.

[12] For simplicity, the previous picture does not depict influences over time. It can be seen as capturing the system over time if the system is taken to be in equilibrium.

[13] There are other causal ways their choices may be informed: they could also get signals of the "type" of legislator by the legislator's public statements, by party affiliation, by interest groups, etc. All of these are instances of the same mechanism.

with the legislature.[14] So long as the constitution of the legislature is periodically determined by the electorate, the electorate has the power to change the decisions and actions of the legislature over time.

Group Action over Time

Now let us connect this back with group action. Intentions and intentional action cannot be understood if we just consider them at a particular moment in time. As Michael Bratman stresses, intentions are part of our system of planning, and are fundamentally future-directed.[15] If we are to make sense of intention, action, planning, reasoning, and responsibility, we need to consider activity over stretches of time.

It is over time that membership conditions are powerful and common constraints on group action. If the legislature is stubborn, the electorate may not be able to change its actions at a particular moment. Over time, however, it can change the constitution of the legislature, ensuring that legislative actions do a reasonably good job at tracking the preferences of the electorate. This is the core of representative democracy: it is not the effect of an electorate on the incentives of legislators, but the diachronic effect of elections on the composition of the legislature.

Mechanisms like these show up in other sorts of groups as well. In the Catholic Church there are conditions for becoming a member of the priesthood or the College of Cardinals. These membership conditions affect the actions of these groups over time. Even membership in the church itself influences action: extreme dissenters can be excommunicated. Similarly for the military. There are conditions for joining, and also conditions on discharge. Sometimes discharge conditions affect the choices made by soldiers, by affecting their incentives. But even when they do not change incentives, they affect the diachronic conduct of the military by getting rid of violators. Likewise in corporations and schools: there are conditions for hiring and admission, and conditions for firing and expulsion. The possibility of expulsion, for instance, affects the incentives of students. But even if those incentives fail, a school simply expels the student. In all of these, both the causal and constitutive mechanisms are at work.

And likewise in sports teams. The success of a sports team is much more dependent on the talent scouts, and by the hiring and firing decisions of

---

[14] Even if we break the causal arrow from the legislative votes to the electoral choices of the population, their choices still partially ground the constitution of the legislature, and hence the legislative votes. For more on this, see Epstein 2008a.

[15] See Bratman 1993, 2014.

coaches, than on investments in training and coaching plays. To see this, one only needs to look at team expenditures: the bulk of spending by any professional team is in fixed contracts to acquire talent. There is little incentive pay, and the coaching staff receives only a fraction of what the players do. Most of what makes a team win is turning over its membership in skillful ways. Membership conditions are not the only constraints on group action over long stretches of time, but they can be the most forceful ones.

What This Case Does Not Do, and What It Does

In the MassDOT/MBTA case, I considered direct constraints on group action. An action at time t, by the MassDOT board, depends on more than the actions of the members. The reason is that external facts set limits on what actions can be taken at t, regardless of what the members do.

In the Microsoft stockholders case, I considered unequal contributions to group action. An action at time t, by the Microsoft stockholder group, depends on more than the actions of the stockholders. The reason is that different members make unequal contributions, and external facts determine which stockholders make those contributions.

The present case is different. It does not show that an action of a legislature *at time t* depends on more than the actions of the members *at that time*. It does, however, show that the electorate figures into determining legislative action *over time*. If elections are frequent enough, or if legislators are sufficiently stubborn, then it can be misleading to focus on modeling the legislators. To model the actions of the legislature over time, what matters may be the preferences of the electorate and the member selection mechanism. This, in fact, may be a direction to investigate for improving models of our current polarized political system.

From the perspective of social ontology, this third category of constraint is only relevant to group action considered over time. It takes on more ontological significance, however, when we move to group intention. As we will see, it is the basis for a robust argument that group intention at t depends on more than the intentions of the group members at t.

Developing a New Perspective on Group Action

It is widely believed that once you have fixed the actions of the members of a group, you have thereby fixed the actions of the group. This is a mistake.

We design groups to serve particular purposes, and do not artificially restrict what pieces of the world should ground facts about them.

Some nonmember grounds are direct grounds of group action and some are more roundabout. We introduce external grounds for group action not just by imposing specific constraints on it, but by making use of hierarchies and membership mechanisms. It is hard to think of a group that is not influenced by one of these sorts of external grounds, if not by all of them.[16]

On their own, each of them is subtle. Add the pieces up, however, and they put group action into a new perspective. Group action is not as general-purpose as individual action. We set up kinds of groups for express purposes. And thousands of years of sociality have endowed us with clever strategies for anchoring supplementary grounds for group action. And thereby for improving the design of groups, helping ensure that they accomplish their purposes. Together, these cases show that member action is just one ingredient in a large list of factors that may ground group action.

Not for a moment should we devalue member action. Of course member action is usually a substantial contributor to group action. But dynamic constraints on group action are a critical tool in our design of groups. And those constraints are often anchored for an express purpose: to bring other facts into the mix.

---

[16] All the examples I have considered in this chapter involve explicit anchors. For many groups, constraints on action are anchored in more complex ways. Family structures, for instance, involve membership conditions and hierarchies of power. These are anchored by historical tokens, practices, environmental facts, and more.

16

Group Intention

Group intentions should be the strongest, easiest case for anthropocentrism about the social world. If any social facts are determined by facts about individuals, it should be facts about the intentions of groups of people. But even these are not. Anthropocentrism about the social world fails even here.

The argument for group intention draws on the last chapter's discussion of group action. But the argument is not identical. For action, we anchor constraints, both direct and indirect. We cannot do quite the same thing for intention. It makes sense, for instance, to enact a law stipulating that such-and-such a group can *act* only under certain conditions. But it does not make sense to enact a law to the effect that a group can *intend* only under certain conditions.

To see how intention is constrained for various kinds of groups, we need to consider how the constraints on action percolate into constraints on intention. And to do this, we need to consider the roles of intention, and what is required for a group to perform these roles.

The fundamental assumption of this argument is one I share with Bratman, and List and Pettit: the model for group practical activity is individual practical activity. Group intentions play an analogous role for groups that individual intentions do for individuals. We figure out the details of systems of practical activity by analyzing the roles of various components of that system for individual people. And then we understand the components of the system of group practical activity to play similar roles.

Our starting points are the same, but I draw a different conclusion. It has to be different, once we take account of the constraints on action I described in the last chapter.

I will start by considering the relation between intention and action, in our system of practical activity. And then turn to the three sets of examples I discussed in the last chapter. For each, I will discuss how a systematic constraint on group action percolates into non-member grounds for group intention.[1]

---

[1] In this book, I remain neutral on debates about the nature of the mental content of

Tying Intention to Action

Group intention plays several roles in a group's system of practical activity. These roles are analogous to the roles individual intention plays for an individual's practical activity. What roles, then, are played by individual intention? Bratman highlights three:[2]

(1) *Intentions set up problems for future reasoning.* When I form an intention to brew a pot of coffee, I do not already need to have worked out the steps for doing so. In intending, I commit myself to an end or aim, which I then need to reason my way through.

(2) *Intentions constrain other intentions.* When I form an intention, it needs to cohere with my other intentions. Suppose I get frustrated as I reason my way through the coffee-making process, and form the intention to smash the coffee pot. That intention conflicts with my coffee-making intention, so I need to reconcile them or abandon one or the other.

(3) *Intentions issue in corresponding endeavoring.* Intentions do not just play a cognitive function, but function in practical activity in the world. When I form an intention to brew a pot of coffee, I go ahead and try to brew a pot of coffee.

In this list, we can see the filaments weaving through different components in the system of practical action. Role 1 pertains to the interaction between intentions, plans, and reasoning. Role 2 identifies a feature of intention in coordinating plans as a whole. And Role 3 pertains to the interaction between intending and acting.

With Bratman, I will understand these to be some of the roles or functions of intentions.[3] Intentions do not need to perform these perfectly, all the time.

---

individuals, in particular the debates between internalism and externalism. If externalism is correct, it is harder to show that group intentions depend on more than the mental states of members. (The externalist takes individual attitudes to supervene on a broader set of factors than the internalist does.) I myself favor externalism, so I take this to be a burden of the argument in this chapter: even granting externalism about the attitudes of group members, group intentions depend on more than even these.

[2] Bratman 1987, 141.

[3] That is, I will roughly understand intention to be a *functional kind*. I will not, however, take a stand here on how exactly we are to cash out this notion. For recent work on functions, see Ariew, Cummins, and Perlman 2002, and Buller 1999. My own view is that functional kinds can best be understood using the anchoring machinery (see Epstein 2014b), but the points I make here about intention are compatible with most if not all theories of functional kinds. In Bratman 2014, 15–18, he gives a somewhat different characterization of roles like these three.

There can be intentions that do not pose problems for future reasoning: I can form an intention and promptly forget it. There can be pairs of intentions that conflict with one another: I might not realize that two of my intentions conflict, or I might notice a conflict but not have figured out how to resolve it. And there can be intentions that do not issue in endeavoring: I can hold onto intentions without ever doing the first thing about them. Roles like these are not criteria for something to be an intention. Intentions normally perform these roles: they do a decent job overall. Particular intentions can malfunction. But intentions are what they are because they are instances of a kind, whose other instances generally do play these roles.

This brief list of roles is only a start. A full account of these three roles would need a lot more detail, nor are these three the only roles of intention.[4] Most of this detail is more than we need, for our purposes. There is, however, one point I need to make about Role 3.

To be sure, endeavoring or trying is closely related to intention. Intentions are not just for deliberating and planning what could be done—they are for issuing in attempts to do things. When I intend to J, I normally try to do J. This is fine, as far as it goes, but it misses a crucial point. Our system of practical activity does not function properly if it issues only in endeavors to do what we intend. It is designed not only to issue in *attempts at action*, but to issue in *successful action*. It is not playing its role if it does not normally issue in successful action.[5]

Take some kind K of agents—a kind like *human* or *circuit court* or *intramural basketball team*. Suppose K has a well-functioning system of practical activity. At any given time, agents of kind K will have various intentions. That is, facts of the following sort will obtain:

At time t_1, agent x intends J_1.

At time t_2, agent y intends J_2.

At time t_3, agent z intends J_3.

If K's system of practical activity is well functioning, that means that another set of facts will also obtain:

---

[4] See Bratman 1987, 140–45, and Bratman 2014, 15–25.

[5] This is different from the much more controversial issue about whether the conclusion of practical reasoning is an attitude or is an action. (See Dancy 2009; Paul 2013; Shah 2008.) I take issuance in successful action to be among the functions of our system of practical activity. But I am neutral on the debate about the ontological dependence of an "unblemished episode of practical reasoning."

At a time after t_1, agent x does J_1.

At a time after t_2, agent y does J_2.

At a time after t_3, agent z does J_3.

Not every intention, for every member of the kind, has to issue in corresponding action at a later time. But normally it does.

This is only a small clarification to the list of roles, but it is essential. It is the reason that constraints on group action matter for intention and other group attitudes. When we anchor constraints on group action, those constraints also affect the grounding conditions for the other components of the group's system of practical activity. Now let us turn to the three kinds of constraints we discussed in connection with group action.

1. Unequal Contributions

Let us return to the Microsoft case. Bill Gates and I make unequal contributions to the action of the Microsoft stockholder group. Now consider some ascriptions of intention to that group:

(16.1) The group of Microsoft stockholders intends to approve the reappointment of board members.

(16.2) The group of Microsoft stockholders intends to accept the proposed acquisition of Nokia.

It is plausible enough that the group can have intentions like these even if some stockholders disagree. The group can have these intentions, for instance, even if some shareholders are intransigent, trying as hard as they can to derail the actions, or reject the bylaws of the corporation, or disagree with the deliberation procedures of the group, or are engaged in bitter litigation against the company.

This assumption is at odds with Bratman's basic analysis of shared intention, and certainly with Easterbrook's. My aim at the moment is not to argue that point. For now, I will leave it at this: let us suppose that these ascriptions are true. If so, then group intentions depend on more than member attitudes.

Consider the relative contributions of different group members to the group intention. Do all the stockholders contribute equally? Or rather, do some member intentions carry more weight than others, in determining the intentions of the group?

To answer these, think about what would happen if everyone's intentions were weighted equally. Suppose the Microsoft shareholders were closely divided about the Nokia acquisition, with many smaller shareholders opposed to it, and a smaller number of large shareholders in favor of it. Given the concentration of ownership in Microsoft stock, suppose the large shareholders have the greater part of the total shares.

If everyone's intentions were weighted equally, then the intentions of the smaller shareholders would outweigh the intention of the large shareholders, with respect to their contributions to the whole. The *vote* would be won by the large shareholders, since altogether they own more stock. But if everyone's intentions were weighted equally, the *group intention* would be the opposite, that of the opponents, since there are more of them.

This is silly. It would mean that group intentions regularly mismatch the actions the group ultimately takes. After all, we have already agreed that some shareholders make a larger contribution to group *action* than others do. If everyone makes exactly the same contribution to group *intention*, then group intention would not normally issue in the group taking the intended action.

For the group's system of practical activity to function well, group intention cannot systematically mismatch group action. And if the intentions of the group of stockholders are to function well—that is, if they are normally to issue in action—that means that the grounds of those intentions need to reflect their stockholdings, just as the action does. Different shareholders contribute unequally to the group's actions, and likewise different shareholders also contribute unequally to its intentions.

This might seem like a small point. But this point alone implies that group intention can depend on more than member attitudes. The reasoning is exactly the same as it was for group action. Different shareholders contribute unequally to the intentions of the shareholder group. The intentions of the Microsoft shareholder group, for instance, are partly grounded by facts like *Bill Gates has more shares than I do*. But that fact, as we have seen, is grounded by facts other than our respective attitudes.

We can depict this with a figure similar to the one for group action (figure 16A). As in figure 15D, facts that depend on more than just the members of the class are highlighted in bold. The group intention does not depend just on group members and their attitudes. It also depends on the facts that ground the assignment of differential powers. Even if we fix the members of the group, and fix their individual attitudes, we still need more to determine the intention of the group.

As I discussed in the last chapter, unequal contributions do not always have this implication. Still, it is common for relative contributions to be grounded by

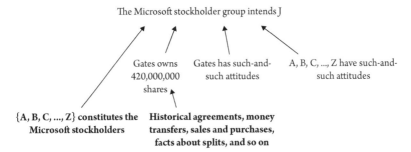

Figure 16A External grounds for members' unequal contribution to group intention

facts apart from member attitudes. In all such cases, group intention depends on more than member attitudes.[6]

A Comment: Intention and Action Do Not Need to Align Perfectly

Group intention fails at its job, if it *grossly* mismatches group action. Still, the match does not have to be perfect. The grounds for group intention do not need to exactly mirror those for group action, even when we explicitly anchor constraints on group action.

After all, there are other parts of our system of practical activity, which intention and action need to harmonize with as well. One role of intention is to issue in action, but intention has other roles too. It still needs to interact appropriately with planning and deliberation. The unequal contributions explicitly anchored for group action might apply nearly unchanged to group intention. It is plausible, for instance, that in grounding the group intention of the Microsoft shareholder group, Bill Gates's intentions have 4,200,000 times the weight of mine, just like his actions do. But it might instead be that intention's other roles distort those contributions.

For example, we might explicitly anchor constraints on planning and deliberation, as well as on action. We might anchor constraints to the effect that all shareholders make equal contributions to certain aspects of shareholder deliberations. And this would temper the inequity among our respective contributions to group intention. Intention needs to harmonize with all of these components.

---

[6] Once again, for the conjunctivist, things are even worse. If we fail to distinguish anchors from grounds, then the facts that anchor the conditions for being Speaker of the House are among the grounds of the intentions of the House.

In short, we cannot just look at the constraints on *action* for groups of kind K, and then deduce the grounds for the intentions of groups of that kind. There is no reason to expect the grounds for group intention to be simple. After all, the grounds of *individual* intention are far from simple: they involve neural states that we have no idea how to disentangle.

Still, we do not need a complete theory of the grounds of group intention, in order to see that group intention depends on *more* than member attitudes. All we need are some unequal contributions to group intention. Bill Gates may not contribute exactly 4,200,000 times what I do to the intention of the shareholder group, but he surely contributes more. That is enough to imply that the group's intentions depend on more than member attitudes.

2. Group Intentions, with Direct Constraints on Group Action

In the second kind of case, we find even more radical mismatches between the intentions of a group and the intentions of its members. It is helpful to illustrate with an exaggerated version, in order to make its characteristics apparent. Consider again the MassDOT/MBTA boards, and the roles of Dorothy and Bertha in determining the actions of the respective groups. External facts like these can have systematic effects on the actions of the groups, even if the members are not aware of them. These effects reverberate into the intentions of the group.

Adding to the Example

Suppose the MassDOT board has a problem with decisiveness. Suppose the members—Alice, Bob, and Carol—have become overwhelmed by their responsibilities. Dorothy notices a daily pattern in the psychology of the board members: in the mornings, they have firm individual intentions to perform some given action. On one given morning they intend to approve a highway bill, and take the vote in favor. By the afternoons, however, they grow tired and confused. The board members start to second-guess themselves, and reverse their earlier intentions and actions. On the given afternoon, they decide instead to reject the highway bill, and take a vote overturning their morning vote. And then they keep waffling: later in the afternoons, they reverse their individual intentions and actions again. And they reverse them again in the evenings. The result is chaos.

So Dorothy decides to remedy the situation. Each day, she gavels the group into session first thing in the morning, and adjourns the group a couple of hours later. In this way, the group is in session while the members are alert and confident. And it is out of session in the afternoons, when they are second-guessing themselves. Because it is only in session in the mornings, the afternoon reversals are impotent. We can even suppose that the group members are unaware of Dorothy's actions. We can stipulate that the psychology of the members follows the same volatile pattern, even after Dorothy starts her decisive gaveling. Despite that, the group is not in session in the afternoons. It is only in session when Dorothy has made the appropriate declarations, and so the actions they take during that time are the only group actions they take at all. Even though they continue to meet later in the day, and attempt other actions, those misfire, as in scenario 2 from the last chapter (figure 15E).

Again, we could tell this story without Dorothy: we could achieve the same result by anchoring the group to be in session in the mornings, and out of session in the afternoons. This would accomplish the same thing. It would render the afternoon activities irrelevant. The net effect is to make the waffling group take consistent action.

The Divergence between Member Intention and Group Intention

Dorothy's actions do not affect the intentions of the individual board members—Alice, Bob, and Carol—over the course of the day. Those three are wafflers, changing their intentions over the course of the day, each day.[7] A theory that takes intention to depend only on member attitudes will thus hold that the MassDOT board has waffling intentions. According to Bratman's theory, for instance, the group intentions flip repeatedly over the course of the day, as the member intentions do.

If the intentions of the MassDOT board were determined only by the attitudes of its members, we would have to abandon the role of group intention in tracking its action. After all, with Dorothy at the gavel, the MassDOT board is taking consistent action. The members continue to waffle, but nonetheless the board acts decisively day after day.

---

[7] As I mentioned in the discussion of group action, it would ruin the supervenience theorist's argument if the theorist took the individual intentions of Alice, Bob, and Carol to be flipped on and off by Dorothy's gaveling. If the individual intentions are merely parasitic on the group intentions, then that theorist no longer has a case for the group intentions to be exhaustively determined by member intentions.

The board takes consistent action over time. It plans, intends, and deliberates its way there. In doing so, the board's intentions have no trouble tracking the board's later actions—so long as facts about Dorothy figure into grounding the board's intentions. In the present case, a systematic external ground is anchored for the *actions* of groups of a given kind. Correspondingly, the grounds for the *intentions* of groups of that kind bend to harmonize with it.

Again, group intention does not need to track group action perfectly. If the parliamentarian were anchored only to have a very occasional effect on the actions of the group, then group intention might still be able to do its job even if the parliamentarian did not figure into the grounds of group intention. Normally, though, we impose external constraints on action so as to have a systematic effect on group action. And where there are *systematic* external constraints on action, those constraints percolate into group intention.

As I pointed out in the last chapter, direct constraints like these are routinely anchored in setting up groups. Group intention is not insulated from this: these anchors also loosen the members' control over the intentions of the group.

3. Group Intentions and Membership Constraints

The final case is a little tricky, but intriguing. To see how it works, we need to think about the system of practical activity over time.

Recall the connection between the roles of intention and action. Let K be a kind of agent with a well-functioning system of practical activity. Then, normally, an intention of an agent of kind K at a given time will line up with the actions of that agent at a later time. If a fact like ***At time t_1, agent x of kind K intends J*** obtains, then so does ***At a time after t_1, x does J***.[8]

Notice that this alignment is not between an intention and a *simultaneous* action. An intention at time t is not typically an intention to act *at time t*. Rather, intentions typically aim at actions at a *later* time. Consider, then, what happens when we anchor a systematic constraint on actions, for agents of kind K. Since intention typically precedes action, it is not only simultaneous intentions that need to harmonize with that constraint. Earlier intentions need to harmonize as well.

---

[8] I say "agent *x* of kind K" rather than just "agent *x*," because this need not hold for *every* agent of a given kind: actions and intentions may only normally line up over the activities of a population.

In the last chapter, I pointed out that membership conditions can have systematic effects on group action over time. They can turn over the membership so as to cut off fringe or dissenting action, or to make the group likely to achieve a certain end. That is, we can use membership conditions to anchor long-term constraints on group action. These conditions affect group action over time. Because of the coordination of action and intention over time, this has a striking consequence for intention. Since group intention at an earlier time t_1 aligns with member action at a later time t_2, that means that group intention at t_1 is not determined merely by member intention at t_1.

As I warned, the point is tricky. To illustrate, let us return to the case of electoral control.

An Electoral Control Example

In modern political systems, it is common for a certain conservatism to be built into the legislative process. Political systems have been expressly designed to slow things down, so that major changes cannot be enacted in a rush. Electoral control is a key reason that this has the desired effect.

Suppose that certain members of the House of Representatives change as they get enmeshed in the politics of Capitol Hill. They arrive in Washington with their idealism intact, keen to promote the interests of their constituents. But the longer they are in town, the more they are pushed and pulled by polarizing think tanks and well-funded ideologues.

To temper phenomena like this, we have made the legislative process cumbersome. A major bill, from initial conception to final passage, takes many legislative sessions. Supposing the electorate is slower to radicalize than the legislators, this allows the electorate to dampen the extremism of the legislature.

Members of the House of Representatives can continue to be susceptible to enthusiasms. They may all, as individuals, have radical intentions at any given time. But the control exercised by the electorate means that radicals are only temporary cogs in a larger legislative machine. The periodic intervention of the electorate moderates the *actions* of the House over time.

The Resulting Effect on Intention

Consider these constraints on action in the context of the House's system of practical activity. Suppose that at the beginning of year 1, the populace elects a set of House members, and the House starts deliberating. By the middle of year 2, the members are succumbing to the pressures of DC politics, and become enthusiastic and faddish. Suppose that the individual members *all* form individual intentions, at that point, to radically overhaul Social Security in some particular way. A process begins, in which they start discussing, deliberating, and planning. Soon, of course, there is an election, which briefly interrupts the process, but the deliberation and planning continue. It takes three more years of wrangling, and then the House (now twice again recomposed) votes at the end of the period.

With all the changes in membership, what is enacted is far from the action the faddish legislators had individually intended in year 2. Instead, the House enacts a modest set of tweaks to Social Security, not the radical overhaul that had set off the process.

And now the crucial question: What was the *intention* of the House back in the middle of year 2? Was it the same as the intentions of the members in year 2? Was it the group intention, in year 2, to radically overhaul Social Security?

The problem again is the connection between intention and action. Presuming that the House has a well-functioning system of practical activity, its intention in year 2 triggered the subsequent process of deliberating, planning, and acting. If the House's intention to do J in year 2 is playing its role, then it should be *that* action—that is, action J—that is taken at the end of the process. But at the end of the process, the House does *not* take the actions that the *members* intended in year 2. It took a different action. So the *group* intention in year 2 had to be different from the member intentions. This, despite the fact that *all* the members had identical individual intentions in year 2.

In short: if the House's system of practical activity is to function well, the intention of the House cannot be the same as the individual members' intentions. Even though the individual members all intended to radically overhaul Social Security, the House could not have intended the same, on pain of breaking the connection between intention and action. And hence breaking the proper functioning of its system of practical activity. Rather, in year 2, the House must have a more moderate intention, with regard to the overhaul of Social Security. Its intention must be determined by more than the year 2 intentions of the members.

Making Sense of This Result

No doubt, this is counterintuitive. Yet if we can free ourselves from over-identifying the group with the members, the point becomes a natural one. Neither the existence of the group nor the constitution of the group is fully determined by facts about the members. Nor is the action of the group. Nor are the other components of the group's system of practical activity. If the group is not identical to its members, then we should not expect that the group intention needs to imitate the member intention, even when all the members have the very same intention as one another.

It might seem that I am suggesting that the intention of the House in year 2 depends on the future, that is, on things that subsequently transpire. But this is not the implication. In a system of practical activity, all the components are built to harmonize with one another in normal circumstances. In the present case, the normal circumstances are ones in which the mechanisms of electoral control temper the radical intentions of the members, moderating them routinely and systematically. The membership constraints on group action are designed to stabilize and restrain action, even against the intentions of members. Their systematic effect is to prevent future action from being capricious. The constraints on group action imposed by the electorate are gradual and predictable. This means that in year 2, we do not need to look into the future to determine how things will normally go. The individual members intend something radical, but the constraints on the House mean that the group intends something along the same lines, but more moderate.

The group's intentions at time t do not need to be determined by events subsequent to t. Rather, group intentions at t are determined by facts allowing those intentions to align, in normal circumstances, with the actions they intend. The intentions of the group, at a *given* time, are partly grounded by the same sorts of thing that constrain the actions of the group *over time*. Namely: the more moderate facts about the electorate.

This vindicates a common feeling that many of us have about groups, but that prevailing approaches do not take seriously. Groups act over long spans of time, through multiple turnovers of their memberships. It is common to have the sense that such groups have intentions that are "bigger" than the members, that is, that groups are pushed and pulled in directions that may conflict with the intentions of the members. We can make perfect sense of this phenomenon without appealing to "group spirits." We anchor systematic constraints on group action, and on other facts about a group. Those constraints cascade into the entire system of the group's activity, including its intentions.

Even group intention is not fully determined by the attitudes of group members. In this case, what is left of anthropocentrism about the social world? Not much. If group intentions do not supervene on member attitudes then, a fortiori, other facts about groups do not. A fortiori, facts about corporations and universities and economies and nations do not. A fortiori, facts about money and boundaries and handicapped parking spaces do not. Building the social world out of people, or modeling by starting with people, is a gross distortion.

Comparing Group Intention to Individual Intention

These conclusions may seem to make group intention very different from individual intention. In some ways it is, and in some ways it isn't.

Our strategy for understanding group intention is no different than the strategy for understanding individual intention. We take both individual and group action to be functionally anchored kinds, and take them to play the same functions in their respective systems of practical activity.

We found that group intention does not decompose in a simple way into the attitudes of members. This makes group intention *more* analogous to individual intention, not less. After all, in analyzing individual intention, we do not break it down into sub-personal intentions of many little homunculi in a person's head. Instead, individual intentions are complex states that normally realize a variety of roles: together with actions, plans, reasoning, and so on, they guide our lives in the world. Similarly for group intention. Group intention has something to do with the individual people that constitute the group, but it does not decompose into them, nor should we expect it to decompose in any simple way.

Still, there are key differences between groups and individuals. Groups have people as members, whereas people do not have people as members. Thus we should expect that the functional role is realized differently for groups than it is for people.

An equally important difference is that we humans are all members of one species, but different kinds of groups are anchored in different ways. This means that constraints on group action can apply to all the actions of all the groups of a given kind, but individual actions are not systematically constrained in this way.

We do not—and cannot—constrain the actions of individuals in the blanket way we constrain the actions of groups of a given kind. Sometimes, we do anchor constraints on individual action. For instance, the same constraints on the actions of the Supreme Court also constrain certain actions of Justice Scalia. He cannot perform certain actions except when the Supreme Court is in session. Similarly, some of my own actions can be performed only in restricted circumstances. For example, I can vote for the president only on certain days. But these sorts of constraints apply to just a few of my actions. My system of practical activity is a general one: I use the same system of intending, planning, deliberating, and acting, for all my practical activity.

Only for a few of our individual actions do we anchor constraints, and those that are constrained tend to be constrained in varying ways. For the kind *human*, therefore, such constraints have a minimal effect on the intertwined functioning of our individual intentions, plans, deliberation, and actions working well together. But groups are different. We anchor constraints on the actions of a particular group, or on the actions of groups of a given kind. Those constraints add grounding conditions to most, if not all, of their actions. Consequently, they have a more significant effect on the functioning of the groups' systems of practical activity.

17

Other Theories I: Social Integrate Models

From the start, I have put forward a flexible conception of *social group*. Groups, as I understand them, are constituted by and only by people. This involves a technical notion—*constitution*—so it is not just an everyday conception. Still, it is meant to be accommodating. It accommodates groups with and without formal structures, large groups and small groups, and groups whose members may be unaware of each other. It accommodates distinct groups that have the same members at a given time. It even accommodates groups that might be memberless at times.

According to certain philosophers, this misses the entire point of social groups. Groups are not just individuals who happen to have some arbitrary property in common. The point of a social group is that it is *social*. And sociality, these philosophers argue, demands more. To be a social group, members must have certain attitudes, beliefs, or commitments, toward one another and toward the group as a whole. To be members of a group, the members must regard themselves as such, in particular ways. It is the integration of members with one another that generates sociality.

This is a widely held contemporary view. Now that we have a good bit of machinery in place, it is easier to explain its shortcomings.

In this chapter, I present the outlines of this "social integrate"[1] model of social groups. I focus mainly on the views of Margaret Gilbert and Michael Bratman. I also say a bit about Raimo Tuomela's approach, but his view is better seen as a mix of this model and the "status" model, which I discuss in the next chapter.[2] Compared to their historical predecessors, the social integrate

---

[1] This is Philip Pettit's term (Pettit 2003), and is also used by Frank Hindriks, in Hindriks 2008.

[2] Pettit is also a prominent representative of this view. But he shares the core of the model with the others, and the points where he differs are somewhat orthogonal to the matters I discuss in this chapter.

theories take a metaphysically audacious stand. They take seriously the notion that groups can act, intend, exercise powers, be subject to norms, bear responsibility, and more. The social integrate theorists make all this sensible and respectable. To criticize these models is not to suggest they are without value. Still, they are built on unfortunate and arbitrary limitations, which makes them a misleading starting point for investigating sociality.

Some Versions of the Model
Margaret Gilbert on Joint Commitment

"A social group's existence," according to Margaret Gilbert, "is basically a matter of the members of a set of people being conscious that they are linked by a special tie."[3] Earlier, I used an example from Gilbert to introduce group intention. If Alice and Bob each have individual intentions to walk to the bank, and are walking side by side, that is not the same as their having the group intention to do so. As Gilbert points out, two individuals walking side by side are not committed to one another: they do not have obligations to one another in virtue of having joined in on a common project. A couple with a group intention, on the other hand, is engaged in a project and is *jointly committed*. It is a social whole in a sense the two individuals are not. The individuals have formed what Gilbert calls a "plural subject."

Social groups, Gilbert argues, are distinctive in having joint commitments like these. Moreover, joint commitments amount to more than just the individual commitments of the members. On her own, Alice can commit herself to walking with Bob. And on his own, Bob can commit himself to walking with Alice. Both can walk together with their individual commitments. And yet even that is not the same as the two having a *joint* commitment. After all, Alice can rescind her individual commitment without communicating it to Bob, and has not violated any obligations toward Bob. When they are jointly committed, however, they can rescind their commitments only together, as a group, if they are not to violate their obligations.[4]

To be a group, in short, requires that the members have "joined forces" with one another, or that they have "pooled their wills." Gilbert lays out a sequence of steps for this, and hence for group formation. To begin with, all the members of the group must be ready and willing to join up with one another. Each of the members on his or her own must be independently ready and willing. (Alice might say, "Would you like to take a walk?" and Bob might think that a walk

---

[3] Gilbert 1989, 148–9.
[4] For more features of joint commitment, see Gilbert 2014, 39–41.

together would be nice.) At that point Alice and Bob are "jointly ready." And once they are jointly ready, Alice and Bob can join forces by overtly expressing their readiness. (Bob says, "Sure, that would be great," and Alice replies, "Excellent, let's go.") That overt expression makes their readiness commonly known, and pools their wills.

Two comments often come up from readers of Gilbert.[5] How can it make sense that a joint commitment is distinct from individual commitments of the members: does this mean that the group is something over and above the members? And how is it possible to generate a joint commitment—which involves a commitment to the group—if the group does not already exist?

Gilbert, I think, can answer these. As I have discussed in previous chapters, many social facts are "initiated." The fact **Clinton was president in 1994**, or the fact **Sarah and I are married in 2014**, are only partially grounded by events in 1994 and in 2014. Most of their grounds are facts about things that happened earlier. They have historical initiation conditions. According to Gilbert, the genesis of a joint commitment, and hence of a group, is similar. It is grounded by historical facts about the participants' "readiness to commit," and also by the overt expression of their readiness. Gilbert, in other words, proposes something like the following principle regarding social groups:

(17.1) It is necessary that: If a set of people are in such-and-such a state of readiness regarding an end (or action, proposition, etc.), and if they all perform such-and-such an overt expression of commitment to it,[6] then those facts ground the following facts:
(1) *The members are jointly committed to the end*
(2) *A new social group (or "plural subject") exists*
(3) *The people in the set are the members of that group*

These conditions are different from ones that would generate individual commitments to the goal. And it is not necessary for the group to exist prior to the "overt expression" that triggers its existence, any more than it is necessary for the marriage to exist prior to the utterances of "I do."[7]

---

[5] See, for instance, Robins 2002; Sheehy 2002.

[6] And if appropriate exit conditions do not obtain.

[7] Tuomela gives a similar overt-expression account in his "bulletin-board view" of joint intention formation. See Tuomela and Miller 2005.

Gilbert's view also accommodates coinciding groups. Suppose three people are jointly committed to painting a house. Then at some point during the day, two of them go off on a walk together. Those two jointly commit to a different end, and so they have formed a different group. But the same can occur if all three of them go off for a walk. The same three people could be members of two distinct groups.[8]

Michael Bratman on Modest Sociality

Where Gilbert takes a firm stand on the nature of social groups and on the essence of sociality, Bratman keeps his ambitions narrower. Instead of making claims about groups in general, he focuses mainly on group actions and intentions. He also imposes a limitation to his analysis:

> The limitation is that my focus will be primarily on the shared intentional activities of small, adult groups in the absence of asymmetric authority relations within those groups, and in which the individuals who are participants remain constant over time. Further, I will bracket complexities introduced by the inclusion of the group within a specific legal institution such as marriage, or incorporation.[9]

These restrictions are substantial. The groups he takes as his paradigm are small. This enables all the members to know one another, to confer, to agree, and to have attitudes toward one another's attitudes. They are composed of adults, so that we can take each member to be an active and full-fledged participant. They lack hierarchy or authority relations, meaning that there is a certain symmetry among the attitudes of all the members. No members have differential responsibility for maintaining the unity of the group. They have unchanging memberships, so there is never a question about what constitutes membership in the group over time. And they are not part of legal institutions, meaning that the group can be seen as an isolated set of individuals, where the basis for group unity is only those individuals.

Bratman's paradigm is similar to Gilbert's. Within this paradigm, they do analyze sociality differently: Bratman is skeptical about Gilbert's claims about the commitments and obligations of group members to one another. He argues that in ordinary cases, Gilbert is right that members of a group have commitments. But he thinks that such commitments can be understood using

---

[8] See Gilbert 1989, 220–21.
[9] Bratman 2014, 7.

more standard notions, like promises and assurances to one another. He also thinks that modest sociality does not *require* joint commitments of Gilbert's sort. Instead, people can form a modestly social group even while freeing one another of such commitments.[10] Gilbert, for her part, denies that group members need all the attitudes Bratman includes. On her view, members of a social group can be jointly committed to an action—and even jointly intend an action—without all having the individual intention that the group perform the action.[11]

The differences between their approaches, however, should not obscure what they share. Both regard sociality as formed by the mutual agreement, attitudes, or commitments of the members of the group. The core paradigm is a set of people all coming together at once, agreeing to be participants, and joining in the common pursuit of an end. The key to sociality is the actors joining forces.

Tuomela on We-Mode Groups

A third notable approach is Raimo Tuomela's theory of sociality. Tuomela builds his analysis in different terms than either Gilbert's commitments or Bratman's meshing attitudes. At the heart of Tuomela's work is the distinction between what he calls the "I-mode" and the "we-mode." Each of us, according to Tuomela, has attitudes in these two modes. Attitudes in the I-mode function in our reasoning for private purposes, and those in the we-mode function in social reasoning.[12] Using the notion of we-mode attitudes, Tuomela builds an intricate account of a group having an attitude *as a group*. For a group to accept, believe, commit, and so on, involves each member having we-mode attitudes that meet a variety of conditions.[13]

Social groups, according to Tuomela, are sets of people who are committed in a certain way to a common ethos. Tuomela understands an *ethos* to be the goals, values, beliefs, and so on, that give the group motivating reasons for its actions. (For instance, the ethos of the stamp collecting club is to facilitate members' stamp collecting.)

Some social groups are *we-mode* social groups. These are the ones that genuinely count as social—that can act, believe, and intend as groups, and so on. We-mode social groups are social integrates, much like Gilbert's and Bratman's. Here is Tuomela's analysis of we-mode social groups:

---

[10] Bratman 2014, 118–20.
[11] Gilbert 2014, 102–6.
[12] Tuomela 2007, 46–64; Tuomela and Miller 1988, 2005.
[13] Tuomela 2007, 65–105.

A collective g consisting of some persons is a we-mode social group if and only if:

(1) g has accepted a certain ethos, E, as a group for itself and is committed to it.
(2) Every member of g as a group member "group-socially" ought to accept E (and accordingly to be committed to it as a group member), at least in part because the group has accepted E as its ethos;
(3) It is a mutual belief in the group that (1) and (2).[14]

In each of these clauses there are several nuances, which may not be readily apparent. I will return to them later on, but the basic idea is this. For a group to be a genuinely social group requires (1) that the group have "collective acceptance" and "commitment" group attitudes toward the ethos; (2) that there are certain norms in place for all the members of the group; and (3) that all the people in the group are in agreement about the program. As I will discuss, Tuomela's view is different from Bratman's and Gilbert's views in an important way. Still, there is a close family resemblance. Tuomela, like the others, regards sociality to involve mutual integration, joining together in pursuit of an agreed end. The members of a group are members in virtue of their all being committed to a common project, not just for themselves, but for the group.

Strategies for Extending the Paradigm to More Realistic Cases

Many people have expressed reservations about this paradigm, on the grounds that it is too idealized and intellectualistic.[15] It is idealized in the sense that it requires a kind of togetherness, solidarity, and unity of purpose that people may seldom exhibit. And it is intellectualistic in that it takes sociality to be built out of fairly complex attitudes by group members. It is possible that these characteristics are not possessed even by the groups that would seem like the best candidates, such as communes, kibbutzim, and jazz ensembles.[16] And it seems unlikely that they are possessed by many groups of central interest to the social sciences—large groups, diverse groups, groups made up of colliding populations, groups with marginalized members, or groups that are created by

---

[14] Tuomela 2007, 19–20. I have left out a number of parenthetical comments.
[15] May 1992; Sheehy 2002; Tollefsen 2002; Wallace 1996, among others.
[16] See Oz 2013; Szwed 2002.

oppression or other external circumstances. The social integrate model seems to demand too much.

Being unrealistic, however, need not be a fatal strike against a model. This is what we do in science and philosophy alike: we hypothesize some core features of the phenomena that interest us, develop simple models of the features, and then try to extend them to further cases.[17]

Gilbert and Tuomela have each gone to some effort to extend the paradigm to more real-world groups. The obvious shortcoming of the model is its egalitarianism—that is, its demand that all members of a social group be full participants in the joint project. To address asymmetric or "structured" groups, Tuomela introduces a distinction between the *operative* and *nonoperative* members of a group. The operative members are members authorized to act for the group. Often, that authorization is assigned by other people who are members of the group but who are not as tightly affiliated. Those other people are the nonoperative members.

Tuomela fills out his analyses of groups, group attitudes, and so on, to accommodate nonoperative members. For a structured group to have an attitude, for instance, only the operative members need to have all the beliefs and commitments that are characteristic of full sociality. Essentially, the operative members are the social core of the group. The nonoperative members need only stand in a weak relation to the operative core.[18]

In order to accommodate structured groups, Tuomela thus adds two elements to his account. One is to distinguish different strengths of mutual integration. Tuomela does not insist that there is just one kind of sociality. Although we-mode social groups exhibit a sort of full sociality, there may be different ways for group members to be integrated with one another, which correspond to lesser forms of sociality. Second, he introduces structured forms of member integration. Instead of requiring that all members of a group be bound to one another symmetrically, some subsets of a group may be integrated tightly, while others may be integrated more loosely.

Gilbert employs a different strategy. Instead of distinguishing two classes of members of a group, she takes a structured group to be established just as unstructured groups are: by having the members form a joint commitment. But in a structured group, the joint commitment they make can be very generic. For instance, all the members might make the joint commitment to allow a few appointed members to plan and act on the group's behalf.

When those appointed members plan and act, according to Gilbert, the group as a whole has a *derived* joint commitment to those plans and actions.

---

[17] Bratman explicitly describes himself as doing this. See Bratman 2014, 7.
[18] Tuomela 2007, 135.

For Gilbert, the subsequent commitment of the group is parasitic on the prior generic commitment. It does not matter, on her account, whether the appointed members meet the conditions of full-fledged sociality among themselves. If the group has committed itself to following Alice's decrees, then the group has a derived commitment to her decrees. Gilbert's approach to structured groups, in other words, retains a kind of egalitarian joint commitment as the key to sociality.

Contrasting the Social Integrate Models with Ours

The social integrate model conflicts in an obvious way with the model we have been developing. I have frequently highlighted facts about groups—including facts about existence, constitution, powers, intentions, and so on—that are grounded by facts unrelated to group members. Such facts may have wildly heterogeneous grounding conditions. The social integrate model demurs: such facts are grounded strictly by the integration of group members. This characteristic remains in Gilbert and Tuomela's extensions of the model, as much as in the basic model. Whether member integration is egalitarian or whether it is structured, in either case they take the sociality of groups to be determined by certain facts about members.

Unfortunately for us, the fact that our model conflicts with the social integrate model does not yet show that theirs is the broken one. After all, our results came out of a broad conception of groups. The social integrate theorists started with a different conception. So it is not surprising that we arrive at different conclusions. To social integrate theorists, most of the groups falling into our broad conception are not genuinely social. From their perspective, the heterogeneity of grounds may merely be a side effect of too broad a conception of groups. We can use the word 'group' however we like, according to social integrate theorists, so long as we recognize that groups on the broad conception do not really exhibit sociality. On their view, the heterogeneity of grounding conditions goes away when we confine ourselves to genuinely social groups.

The social integrate theory analyzes sociality in tandem with developing a conception of genuinely social groups. We have followed a different strategy: we started from a broad conception of groups, independent of an analysis of sociality. We developed machinery to approach these entities, and then used it to work through the grounds of various social properties.

Under our broad conception of groups, many groups are entirely uninteresting. By no means do they act or intend, are they jointly committed, or are they otherwise interestingly social. We did not start with agency or sociality, and then develop our theory of groups on that basis. Instead, we started with a loose notion, and then investigated what it takes for a group to have certain interesting and distinctive properties.

So, the two strategies are different. Why reject the social integrate approach? Because at its heart lies a damaging structural assumption.

The Structural Shortcoming to the Social Integrate Models

Back in chapter 11, we kicked off our work on the grounding of facts about groups. Even before making use of any features of groups, or proposing grounding conditions for these facts, we brainstormed a long and diverse list of such facts. Subsequently, we embarked on working out their grounding conditions. From the start, it was clear that various facts are grounded very differently from one another, even when we limit our focus to a single kind of group. In the succeeding chapters, we found that even subtle differences between two facts can mean big differences among the facts that ground them. Recall, from chapter 13, how simple it is to ground this fact:

(13.1) *One new intramural basketball team comes to exist at time t,*

but how complex to ground:

(13.2) *An intramural basketball team exists at time t.*

It is not just that grounding conditions for these involve more than facts about the members. Even more fundamental is that there are many facts about groups, basic and complex, each of which may have different grounding conditions.

Consider, however, formula (17.1), where we gave a rough summary of Gilbert's analysis. This formula looks familiar: I have basically written it in the form of a frame principle. We can understand Gilbert as putting forward a set of grounding conditions for certain social facts. Yet there is a key difference between (17.1) and all the frame principles we discussed in earlier chapters. The difference is not in the content of the grounding conditions, but rather, in what they ground: that is, facts (1), (2), and (3).

In all the earlier frame principles, we have given the grounding conditions for *one* fact. But formula (17.1) does not do this. Instead, it gives one single set of grounding conditions for many. Once a set of people satisfies the conditions for sociality, that does not just trigger the existence of a social group. It also triggers the fact that the group has the constitution it does. And it attaches norms to that group. Gilbert's account implicitly assumes that these various facts about a social group are all grounded by the same thing, all at once.

In our inquiry, we were forced to develop the machinery to accommodate distinct grounding conditions for various facts. Starting with the narrow paradigm, in contrast, leads Gilbert to see sociality as one unified cluster of properties, triggered all at once by the formation of a joint commitment. So long as she sticks to the social integrate paradigm, she is not forced to disaggregate the grounding conditions for the different facts in that cluster. But the paradigm is a special case. Only for an unusual kind of group are all these different facts grounded in the exact same way.

This characteristic is common to the variants of the social integrate model. The idea behind the model is to look for the grounds of sociality, in general, in the integration of group members. What explains the existence and unity of a social group is that the members join forces, in one way or other. A similar assumption is built into Bratman's account, for instance, even though he goes out on less of an ontological limb than Gilbert does. In Bratman's theory, group intention is explained by one complex set of conditions being satisfied by the members of the group. Since Bratman does not make claims about the nature of groups in general, he leaves open the possibility that there are other ways social groups can be formed. But his account does not introduce any resources beyond the integration of member attitudes.

Behind the Special Case

It is no accident that the social integrate model focuses on small, adult, and unchanging groups. This narrow focus and the structural assumption behind the model are two sides of the same coin. The model takes member integration to be the trigger for sociality, one kind of ground for a variety of facts about groups. This means that the model can only make decent sense of one kind of group: groups that plausibly have one single set of grounds for several facts about them. The basic architecture of the social integrate view does not just stake a claim about the grounds of sociality. It rules out entities that have different grounding conditions for facts about their existence, constitution, norms, and so on.

This, most likely, is why theorists assume that all these facts are grounded by facts about the members. Despite the metaphysical advances made by the social integrate theories, these theories continue to identify groups too closely with their constitutions. They agree that a group can be treated as an independent entity, with properties diverging from the properties of its members. But they cannot bring themselves to pull groups any further away from their constitutions. Instead, they assume that the thing that constitutes a group must do all the explanatory work about properties of the group.

Social integrate theorists, in a sense, see social groups as if they were sophisticated ant colonies. This might seem like an odd thing to say. After all, their model builds sociality out of attitudes and commitments of members, and ants do not have any of these. Nonetheless, these theorists still regard social groups as exhausted by their constitutions not so differently from ant colonies. The social integrate model is anthropocentric in the same way an ant colony model would be ant-o-centric. It is true that these theorists take social groups to be grounded by *intentional* facts about members—facts about their attitudes and norms—while ant colonies emerge from nonintentional facts. But intentional facts remain merely facts about the constituting membership, albeit complicated ones. The social integrate model underestimates the difference between the objects of the social sciences and the objects of the natural sciences: most of the properties of a typical natural object are inherited from its constitution, but this is not so for social objects.

What Social Integrate Models Cannot Do

The social integrate models do not only have theoretical flaws. Their limitations show themselves even in minor departures from the small-group paradigm. Membership changes, for instance, are awkward to deal with on an account like Gilbert's. Suppose that Alice and Bob and Carol jointly commit to paint a house, and go to work on it. Then Dave comes along and joins the group as well. It is plausible enough, as Gilbert holds, that Dave "pools his will" with the others.[19] But when he does this, what happens to the group? Does the old group go out of existence and a new one come into existence? If not, then why does the initial formation of the joint commitment create a group, and not the second?

A related problem comes up for coinciding groups. I mentioned above that Gilbert allows that distinct groups may have the same membership at the same time. For instance, the group that paints a house may be distinct from the group that goes for a walk, even though the groups have the same members. But this too raises a problem for the identity of groups. When is a new joint

---

[19] Gilbert 1989, 220.

commitment merely an added commitment to an existing group, and when does it issue in a distinct group?

There is, of course, a straightforward approach to answer questions like these. That is the one we took in chapters 11, 12, and 13, where we built on the divergent grounding conditions for facts about existence and constitution. But this is not available to Gilbert's model, which takes the grounds of joint commitment as the common trigger for both.

The only resources Gilbert has, in order to address questions like these about group identity, are variations in the joint commitments of the members. To explain why one new commitment gets added to the same group, while another new commitment creates a new group, she appeals to differences in the member commitments, and in the grounds for those commitments.[20] Yet these resources are too limited. There is more to the identity and persistence of groups than the commitments made by their members. We might have one group that has both the commitment to paint a house and to walk together, and we might have two distinct groups each having one of those ends. If the identity conditions for a group are not fully determined by member commitments, then it does not matter how nuanced a taxonomy of commitments we develop. Commitments will not suffice.

Social integrate models also hit roadblocks when they are applied to cases with internal hierarchies or power differentials. Already with the example of Mrs. O'Leary's class, the social integrate model runs into trouble. The model needs to account for the fact that the class president has more voting power than the other students. To do so, the social integrate model would look to various attitudes or commitments among the students. But it is not clear that these are or need to be present: it was Mrs. O'Leary, not the class members, who put those powers in place. If the class members do not have the requisite attitudes and commitments, then we need to turn to an extension strategy, such as Tuomela's operative/nonoperative extension or Gilbert's derived commitments. But even for such a simple case, it is not obvious how this should go. What is the larger group in which the operative students are embedded? What is the larger group that has derived commitments to the decisions of the students?

We might be able to piece together a social integrate model for this case, but it seems like forcing a square peg into a round hole. The case is simple to explain. We did it with one diagram: figure 15A. And we were able to draw an equally simple diagram for the case of the Microsoft stockholders (figure 15C). That case is even more trouble for the social integrate model: in the Microsoft

---

[20] Gilbert 1989, 220.

case, it does not matter how many people we include in the narrative. Attitudes and commitments will never be adequate to account for the power differentials of stockholders, in acting and intending.

Sociality and Group Agency

It would seem that the social integrate theorist has a trump card: group agency. The social integrate theorist argues that groups in the broad conception do not exhibit genuine sociality. And if there is any crucial characteristic of "genuine sociality," it is that the group has group intentions, takes group actions, goes through group deliberations, makes group plans, and so on.

As I acknowledged, many broad-conception groups are not plausibly agents in this sense. However, in the last couple of chapters—working with the broad conception—we have found that group agency is realized by different kinds of groups in rather different ways. If a group of a given kind has sophisticated membership conditions, it may realize a system of practical activity even if the member attitudes alone would not suffice. Similarly for groups anchored to have hierarchical powers, or divisions of labor, or sensitive existence conditions, or activation conditions. Various kinds of groups realize the elements of agency in different ways: there is no common signature to group agents.

This result should not be particularly surprising, given a broadly functional understanding of intention, action, and agency as a whole, as it applies to groups. As I discussed in chapter 2, functional properties are often "multiply realizable": they can be realized in different ways, in different kinds of objects. When we restrict the domain to small, adult, unchanging groups, there may only be one general way a system of practical agency is realized. But when we loosen these restrictions and consider broader varieties, we find that it is realized differently. That is not to devalue the detailed inquiry into more restrictive cases, such as the small-group paradigm. It is helpful to see how a system of practical activity can even in principle be realized by even one kind of group. But we have to be careful not to lose sight of the fact that the social integrate paradigm only gives a single snapshot. From just one special sort of realization, we cannot derive a common set of grounding conditions for group agency.

The social integrate model looks for all group agency under the streetlamp of integration among the members. But by ignoring nonmember grounds, it fails to see that member integration alone is neither necessary nor sufficient for group agency. If certain anchors are in place, and if external conditions are felicitous, then even a poorly integrated group can realize a system of practical activity. And if other anchors are in place, and if external conditions are infelicitous, then even a well-integrated group can fail to be an agent. Restricting the paradigm not only overlooks group agents outside that paradigm, but also leads theorists to look for conditions for agency where they cannot be found.

18

Other Theories II: Status Models

In recent chapters, we have barely mentioned either anchoring or Searle's theory of institutional facts. We have focused on how facts about groups are grounded, not how their grounding conditions are anchored. Anchoring came up only when we were thinking about group intentions and functional systems of practical activity. And Searle came up only in a side-note to chapter 11, where we observed that his constitutive rules conflate the grounding conditions for facts about existence, constitution, and more.

There are a couple of reasons, however, that a theorist might take a Searle-like story to be pertinent to social groups. One is all the talk of collective attitudes. Collective attitudes are closely related to sociality, according to the social integrate theories we examined in the last chapter. But we have also seen collective attitudes before, in Searle's argument that constitutive rules are put in place by collective acceptance. Now that we are considering collective attitudes at the center of sociality, it is natural to wonder if there is some relation between those attitudes and the collective attitudes in Searle's account.

A second reason is that some groups seem to fit better into a Searle-like story than into a social integrate one. We noticed in the last chapter that the social integrate model does not work well for many groups. Some groups, like courts and legislatures, seem to be able to intend, act, and so on, despite poor integration among the members. In earlier chapters, I explained how this can occur: for a group to be an agent, features of it need to realize a system of practical activity, and it is not only properties of members that figure into this realization. But maybe we should consider an alternative strategy: namely, some groups are agents because we assign them the *status* of being agents, rather than because they have the right structure. This is roughly Searle's approach. Recall Searle's boundary example: a boundary is created by our collective acceptance that a line of stones has the status of functioning as a physical wall. To do that, the line of stones does not need to have the physical properties of a wall. The stones are just a substrate onto which we project a status. Similarly, perhaps we assign some groups the status of functioning as agents, even though they do not by themselves meet the conditions that group agency requires.

I hinted in the last chapter that Tuomela's approach to social groups was different from Gilbert's and Bratman's approaches, despite appearances. Even though elements of Tuomela's theory resemble the social integrate theories, he also sees sociality through the lens of Searle's theory. Social groups, according to Tuomela, involve status assignments to collections of people.[1]

Frank Hindriks, in a series of recent papers, also takes a Searle-style approach to social groups. But Hindriks makes a sharper point: he argues that there are actually two categories of groups. One is the social integrates, and for these, the theories discussed in the last chapter are roughly correct. But there is another category that Hindriks calls the "corporate agents." According to Hindriks, these include many of the examples I have been discussing in this part of the book, such as courts and legislatures, and also such things as corporations and universities. Corporate agents, in Hindriks's view, are the products of status assignments.

The status model is tempting, especially once we notice the inadequacies of the social integrate model. It draws—as it should—on a broader set of examples of social groups. Moreover, it notices that people play a role in anchoring certain facts about groups, not just in grounding them. Despite these positive impulses, however, I argue in this chapter that the status model heads down the wrong track. It does not really make sense to say that we assign to a collective the status *being an agent, having an intention,* or *taking an action*. Even "corporate agents" need to realize systems of practical activity, in order to plan, act, or have intentions. Of course, we do anchor certain frame principles that bear on these. People play a role in anchoring powers, rights, obligations, existence conditions, membership conditions, activation conditions, etc. But this is not enough to make something an agent. When we anchor these in sophisticated ways, there may be minimal burden on group members, for the group as a whole to perform the functions of practical activity. But in no case does it make sense to see *agency* or *the ability to act* as assigned, authorized, or projected.

I will briefly characterize Tuomela's and Hindriks's versions of the status model, and then challenge the approach. Subsequently, I return full circle to an issue I discussed near the beginning of the book: the distinction between "Type 1" and "Type 2" entities.

---

[1] Frank Hindriks classifies Tuomela differently: he takes Tuomela's account of sociality to be representative of the social integrate model (see Hindriks 2008, 125–6). I think it is more accurate to regard Tuomela's theory as a hybrid, but the difference in our readings is largely a matter of emphasis.

Tuomela on Status Assignments

In the last chapter, I described Tuomela's view in the context of the social integrate model. The first condition for a collective to be a we-mode social group is this:

(1) *g* has accepted a certain ethos, E, as a group for itself and is committed to it.[2]

On the surface, this resembles Gilbert and Bratman. It looks like the conjunction of a Bratman-like claim—the group has a collective acceptance attitude toward E—and a Gilbert-like claim—the members are jointly committed to E. But Tuomela goes on to say more:

> Collective acceptance here conceptually requires reflexive acceptance: necessarily, the members collectively accept E as *g*'s ethos if and only if E is *g*'s ethos. Given this, our account entails that a we-mode group is a collective artifact and indeed an organized institutional entity.[3]

This passage needs to be unpacked a bit. What is *reflexive* about the group's accepting an ethos? How are we to understand *the members collectively accept E as g's ethos if and only if E is g's ethos*? And what does Tuomela mean in saying that social groups are *collective artifacts* and *organized institutional entities*?

Elsewhere, Tuomela fills in the gaps. When the members collectively accept E as *g*'s ethos, according to Tuomela, they are not accepting it as individuals for themselves. Rather, they are accepting it *as a group for itself*. To use Tuomela's coinage, their acceptance has "for-groupness."[4] Consider, for instance, a stamp-collecting club. The group collectively accepts, as its ethos, facilitating members' stamp collecting. According to Tuomela, this means the following: the members of the group accept that the group counts as an entity having the function of facilitating members' stamp collecting.

With the terms 'collective artifact' and 'organized institutional entity', Tuomela is alluding to his refinement of Searle's theory of institutions.[5] Money, in Searle's view, is an organized institutional entity: it is created by our collective acceptance of a rule assigning to a commodity the status of having

---

[2] Tuomela 2007, 19.
[3] Tuomela 2007, 20. I have omitted Tuomela's parenthetical comments for readability.
[4] Tuomela contrasts for-groupness with I-mode progroupness in Tuomela 2007, 52–7.
[5] Tuomela 1995, 438; Tuomela 2007, chap. 8.

a monetary function. Likewise for groups. When a group collectively accepts an ethos, it accepts it *as a group for itself*, which in turn involves the members of the group assigning themselves the status of being a group. This is the sense in which this collective acceptance, for Tuomela, involves reflexivity: it involves the assignment of group status by the members to themselves.[6]

Clarifying Tuomela's Distinctions

Tuomela's basic case involves reflexivity in his sense. But if sociality is a matter of assigning a particular status to a collection of people, then we can also assign that status to others. In the last chapter, I mentioned Tuomela's distinction between the "operative" and "nonoperative" members of a group. In a structured group, according to Tuomela, the nonoperative members authorize the operative ones to act on the group's behalf. Now we can be more specific about how Tuomela understands authorization: it involves the assignment of a status to a collection of people.[7] In an unstructured group, the members assign a status to themselves. In a structured group, the nonoperative members assign a status to the operative members. The status, that is, of taking action on the group's behalf. Both structured and unstructured groups involve status assignments, but in the structured group, a status is assigned to a subset of the members.

Another variation is that statuses like these do not need to be assigned to a *specific* set of people. Instead, they can involve the assignment of tasks and rights to *sets of roles*, which people can fill. Just as a set of people can authorize Alice, Bob, and Carol as an organized institutional entity, they can also authorize a set of roles *x, y,* and *z*—roles to be filled by people—as an organized institutional entity. Alice, Bob, and Carol might initially fill those roles, but then Dave might replace Carol, while the same institutional entity persists. This is what Tuomela calls an "organization."[8]

Tuomela puts status assignments at the core of sociality, and insists on the "artifactual" and "institutional" character of we-mode social groups. Still, it seems fair to regard his model as a hybrid: a combination of the social integrate model and a Searle-like model. For instance, in a structured group, the "nonoperatives" assign a status to the "operatives." But that does not lighten the

---

[6] Notice that this is not the same notion of reflexivity that other theorists employ. For instance, Guala's notion of reflexivity (see chapter 4) is much weaker than Tuomela's.
[7] For discussion of authorization, see Tuomela 1995, chaps. 4–7; Tuomela 2007, 129–34; Tuomela 2013, 160–72.
[8] For detailed discussion of organizations, see Tuomela 2002, 186–92.

conditions on the social integration of the operatives. The operative members, in Tuomela's account, need to meet all the requirements for we-mode sociality.

Thus as I commented in the last chapter, Tuomela's theory does not make much progress accommodating groups that fail to meet the requirements of the pure social integrate model. In a sense, Tuomela's model merely raises the threshold for sociality: we do not just need collective acceptance or commitment to *something* (as Bratman and Gilbert hold) but rather we need a very specific kind of collective acceptance and commitment.

Hindriks on Corporate Agents

In contrast to Tuomela's hybrid theory, Frank Hindriks proposes a more straightforward status account. Hindriks divides group agents into two categories: the social integrates and the corporate agents. Social integrates are the types of social groups I discussed in the last chapter. They have an internal structure, a kind of integration among the members of the group, in virtue of which they are group agents. Social integrates are collective agents from an *internal perspective*.

Corporate agents, on the other hand, have agency in virtue of being assigned a certain status. They are collective agents from an *external perspective*.[9] The members of a corporate agent need not have the degree of internal integration that the members of a social integrate do. What matters instead is that they are assigned the appropriate sort of status by outsiders.

Hindriks's treatment of status assignments is a modification of Searle's. Searle gives one constitutive rule for a variety of different facts—facts about existence, constitution, powers, obligations, and so on. Hindriks cleaves off what he calls "status rules" from the constitutive rules. He interprets constitutive rules more narrowly than Searle does: Hindriks's constitutive rules only give the conditions for an entity to be a corporate agent of some kind. They do not assign powers, obligations, and other normative attributes to corporate agents of that kind. Instead, it is the status rules that do this.[10] Thus we might have a constitutive rule such as *{Alice, Bob, Carol} is a corporate agent of type K*, and a status rule such as *Corporate agents of type K have the right to award diplomas*, or *Corporate agents of type K are able and obliged to file quarterly reports with the Securities and Exchange Commission*. Hindriks agrees with Searle as to

---

[9] Hindriks 2008, 119.

[10] Hindriks 2008, 130–34; Hindriks 2012, 98–103. Despite this separation, a constitutive rule and a status rule do basically the same work together that Searle's constitutive rule does. Hindriks fails to see that different facts involve different grounding conditions.

how these rules are put in place: namely, by the collective acceptance of members of the community.[11]

The term 'corporate agent', of course, suggests that corporations are the key paradigm of such things. And indeed, the central examples Hindriks discusses are limited liability corporations and universities. My own view is that corporations and universities are not the best cases to choose. It is often assumed that they are constituted by and only by people.[12] But this is unlikely. Corporations and universities bear a variety of constitutive relations to material things—people, assets, inventory, buildings, property, equipment, and more. Consequently, I will continue to leave them aside. Still, Hindriks would consider many of the groups I have discussed in earlier chapters—the Supreme Court, legislatures, intramural basketball teams, and so on—to be corporate agents as well.

The actions of a corporate agent, according to Hindriks, are performed by some of its members exercising the powers they have, in virtue of being members. Members can also authorize a nonmember to act on the corporate agent's behalf. A corporate agent, for instance, can authorize a lawyer to file a lawsuit on its behalf, and in doing so, the corporate agent files suit. Such action, according to Hindriks, counts as an action of the corporate agent because the members are exercising their powers in re-assigning a power to the nonmember.[13]

It is possible for a corporate agent also to be a social integrate. A university faculty, for instance, may be both.[14] It is assigned a status externally, by administrators, trustees, students, and so on. And it also may conform to the conditions for a social integrate.[15] Likewise, it is possible for a social integrate also to impose a status on itself reflexively, along the lines of Tuomela's basic case. A church hierarchy, Hindriks proposes, accepts status rules for the church members themselves. But this is not the norm for social integrates: Hindriks does not hold that ordinary social integrates involve these sorts of status assignments. And in any case, if a group assigns itself a status and the norms are not accepted by the society more widely, then they are not norms for anyone but those members.[16]

---

[11] Hindriks 2008, 131.
[12] Hindriks too assumes this: see Hindriks 2013, 419ff.
[13] Hindriks 2013, 420.
[14] Perhaps this is bad example—university faculties tend to be about as fractious as groups can get.
[15] Hindriks 2008, 126–7 also proposes his own variant on the social integrate model.
[16] See Hindriks 2008, 140.

Status and Agency

The motivation for the status model is clear: many groups do not come close to meeting the conditions for social integrates. Yet they are unified and exercise agency. If the source of this is not internal integration, external assignment seems like the natural alternative.

However, being an agent is not plausibly a matter of having an assigned status. To see the problem, consider how status assignment works.

Back to the Basic Case

Consider again Searle's "boundary" case. In his basic example, we collectively accept that a particular line of stones has the status of being a boundary, that is, of having the function of a physical wall. Suppose that case works just as Searle says it does.

Now, what happens if we assign *that* status to a set of people, rather than to a line of stones? We collectively accept that a particular set of people—{Alice, Bob, Carol}—has the function of a physical wall. We have assigned them a power: where they stand, villagers are supposed to keep in, or keep out, or pay a toll. They stand in a row, and hence exercise this power.

Are they plausibly a group agent? After all, they are a set of people to which a power has been assigned. But if that is what it takes to be a group agent, then what about the line of stones? Is it a group agent too? Presumably not. But then, what is the difference, in the status theory, between people as a substrate and stones as a substrate?

Maybe the issue with this example is that we should not regard a line of dormant stones or of standing people as *exercising powers*. So consider a more active case. Suppose we assign a power to a pendulum—the power to marry a couple when it swings to the right, and to divorce the couple when it swings to the left. Or suppose we assign a status to a bottle, as in the game of "spin the bottle": the direction the bottle is pointing indicates who is owed a kiss. And then, instead of a pendulum or a bottle, suppose we assign one of those statuses to {Alice, Bob, Carol}. Suppose, for instance, that the person at whom Alice, Bob, and Carol are all looking is the person owed a kiss.

According to the status account, {Alice, Bob, Carol} is a corporate agent in such a case, exercising its agency by looking at a person. But if so, then what about the pendulum or the bottle? After all, they have the same powers, and in virtue of the same thing: our collective acceptance that they do. When Alice, Bob, and Carol look at someone, they have exercised their assigned powers and hence carried out an action as a corporate agent. Is the bottle also an agent, in its exercise of assigned powers?

Where the Status Theory Is Led

The status theorist might respond in one of two ways. Perhaps there is something defective, in the above examples, about the powers or statuses being assigned. Perhaps only certain powers, rights, responsibilities, and so on, suffice to make a collection of people a corporate agent.

But it is not clear what those are, or what the reason would be for such limits. And even if we did find such limits, then we could conversely assign *those* powers to a line of stones or a pendulum or a spinning bottle. And we would have the same problem.

A different and better response by the status theorist is this: it matters what the *substrate* is, onto which a status is assigned. Maybe collections of inanimate objects, regardless of the status assigned to them, are not agents. The grounds for a set of people to constitute a corporate agent include facts about them, or perhaps the fact that they are people. Corporate agency cannot just be assigned to an arbitrary substrate.

This seems right, but when we head down this road, we start to see that there must be *many* grounding conditions for corporate agency. Presumably, the reason corporate agents must be constituted by people, not stones, is that people can act intentionally. Corporate agency is not just grounded by arbitrary facts about a substrate, but by the intentional actions of people. Yet even this is not enough: it cannot be sufficient for a corporate action to be triggered by an *arbitrary* intentional action by a person.

For instance, suppose we assign a power to {Alice, Bob, Carol} in the following way: when Alice brushes her teeth, that counts as the exercise of some completely unrelated power. We collectively accept, for instance, that her brushing her teeth is the exercise of the group's power to endorse a candidate for president.

Though brushing her teeth is an intentional action by Alice, this case is little different from the pendulum and bottle cases. When a group agent performs some action, more is needed than just that the members are taking *some* intentional action, which we have arbitrarily assigned to be the performance of an action by the group.

In short, a faithfully Searle-style approach does not begin to do justice to agency. Projection by collective acceptance cannot be enough to yield an agent. Implicitly, Tuomela and Hindriks recognize this: they assemble much more than a strictly Searle-style picture would entail. Tuomela, as I discussed, requires the operative members of a group—those who perform the actions on behalf of the group—to meet all the requirements for we-mode sociality.[17] Hindriks holds that group action needs to be *enacted* by group members, or by an agent authorized by members of the group, though he does not go into detail on what it takes for this enactment to occur.[18]

How, then, should we fill out the requirements for group action? This is something we have already covered: we sketched it back in chapters 15 and 16. What the grounding conditions on group action are anchored to be depends on how groups of that kind realize a system of practical activity. In certain cases, the actions of a group may be fully grounded by certain actions of a particular person, under certain conditions. But when that does occur, it is a consequence of the way the group realizes an overall system of practical activity, and how that person's actions fit into it. And if a group does realize a system of practical activity, an additional status assignment is not needed, nor does it add anything.

We can anchor frame principles that affect our system of practical activity. We can anchor hierarchies, jurisdictions, deliberation procedures, membership conditions, assignments of power, and so on. The status theorists, however, mistakenly interpret these to be assignments of agency. They are correct to observe that the social integrate model does not adequately accommodate these, but they misdiagnose what it takes to anchor the frame principles for group agency.

There are many kinds of groups, but there are not multiple kinds of agency—agency that is a product of the internal structure of the members, and agency that is a product of the external assignment of status to the members. Instead, group agency is one thing: it is a matter of the group satisfying the grounding conditions that are anchored by the way groups of that kind realize a system of practical activity. At the same time, facts about a group are not determined just by facts about its members. So the realization of that system can look very different from one kind of group to another.

Revisiting the Type 1/Type 2 Distinction

The division of groups into the social integrates and the corporate agents recalls the distinction between "Type 1" and "Type 2" objects from chapter 6.

---

[17] See Tuomela 2007, 112–14, and Tuomela 2007, 132–3.
[18] Hindriks 2013, 420. Hindriks's view may be that an agent enacts a group action when the agent acts in accordance with a collective decision procedure assigned to the agents. This moves in the right direction, but even if we anchor aspects of a group's decision procedure, that remains just one part of group's system of practical activity, with which group action needs to coordinate.

(Examples of Type 1 included a mob, a flow of commuters, and the Jewish people; those of Type 2 included a handicapped parking spot, a stuffed animal tea party, and an unkosher animal.)

The social integrate view seems to take groups to be Type 1 entities. Not every Type 1 entity counts as a social group, according to the social integrate view—a mob, for instance, may not be integrated in the requisite ways. Still, these views conceive of the sociality of the social group as being "built out of" certain properties of its members, together with relations among them.

In contrast, the status model seems to take social groups to be Type 2 entities. They are formed in the Searlean style, with the imposition of a status on a substrate. On its surface, the Supreme Court might not appear to be like an unkosher animal. But according to the status model, that is just because the substrate is different: it involves a status imposed on a collection of people, rather than on a lobster or a pig.

After all our examination of grounding and anchoring, this might seem like a good distinction. To paraphrase Shakespeare: some collections of people are born social groups or achieve sociality, while others have sociality thrust upon them. However, we should also notice that the Type 1/Type 2 distinction was a temporary one, which we used and then discarded. We used it to clarify the difference between anchoring and grounding, and to distinguish two mistaken theses about how individuals "make" the social world—ontological individualism and anchor individualism.[19] The Type 1 examples were the ones theorists use to defend ontological individualism, and the Type 2 examples to defend anchor individualism. As we have investigated groups and seen the appeal of ontological individualism vanish, however, the Type 1/Type 2 distinction has faded as well.

The intuitive Type 1/Type 2 distinction was between entities that are "built" out of individual people, and entities that have social properties projected on them by people. In this part of the book, however, we have seen these characteristics cut across one another. Even when we stick to entities that are constituted by and only by people, there are countless ways of filling out the following table (table 18a).

---

[19] I have not, of course, directly challenged anchor individualism in this book, but I do discuss it in Epstein 2014b.

Table 18a **A variety of facts, grounding conditions, and anchors**

Fact about group Y	Grounding conditions for the fact	Anchors for that fact's frame principle
1. Y is constituted by X		
2. Y exists		
3. Y has such-and-such activation conditions		
4. Y has such-and-such powers		
5. Y has such-and-such rights		
6. Etc.		

The intuitive distinction between Type 1 and Type 2 entities, in other words, arises from a faulty assumption about the relation between entities and their constitutions. The social properties of Type 1 entities, it is assumed, are determined exclusively by facts about their constitutions. Type 2 entities, of course, cannot possibly have their social properties determined by their constitutions, since they are just constituted by things like areas of pavement, arrangements of stuffed animals, and real animals. So they must acquire their social properties in a different way: namely, by being projected onto physical substrates. However, it is a mistake to think that facts about the paradigmatic Type 1 entities depend so much on their constitutions. Once we see this, the contrast between them and the paradigmatic Type 2 entities largely goes away.

We can, of course, categorize groups in various ways. But at least until we have made more ontological progress, we should take intuitive taxonomies with a grain of salt, or else we end up reinforcing artificial divisions and creating unnecessary puzzles. If we assume that corporate agents are created by assigning statuses to substrates, that immediately raises the question, *what are the substrates?*[20] And then once we identify various substrates, it is natural to classify group-type according to substrate-type. It is natural, for instance, to distinguish groups that involve a status assignment to a collection of people

---

[20] The ongoing debates in the literature about "freestanding Y-terms" in Searle are an artifact of the same assumption. See Hindriks 2013; Searle 2008, 2010; Smith 2003, 2008; Tuomela 2011.

from those that involve a status assignment to a set of abstract roles, which people can then fill in.

Our more general treatment of frame principles dispenses with the idea of a status projected onto a substrate altogether. We anchor grounding conditions for facts about a group's constitution. Depending on the grounding conditions, they might imply that the group has a fixed membership. But there is nothing more basic about a fixed membership than a changing membership, nor is there a need for an ontology of "roles" as separate from an ontology of sets or collections of people.

Moreover, how a group is constituted is just one of many sorts of basic facts about it. We can, of course, classify groups along whatever dimensions we like. But too much attention has been paid, in too many theories of groups, to group members and to the human factors that figure into making them members. These are only a small piece of the story.

Looking Ahead

Frontispiece to Thomas Hobbes, *Leviathan*, 1651.

The famous frontispiece of Thomas Hobbes's *Leviathan* depicts the body of the sovereign composed of people. In the 350 years since its publication, we have made good progress in understanding the nature of the state. The state is not an organism, with limbs subordinate to a unified whole. Effective state authority does not require an absolute monarch. The leader of a state need not have religious authority. To bring Hobbes's picture up to date, we would remove the crown, lop off the head, and dispose of the crosier in the left hand, if not the sword in the right.

The center of the drawing, however, would remain untouched. In the prevailing contemporary view, society is a composite of interacting people. In this respect, Hobbes's ontology retains its grip.

Perhaps this situation is not surprising. Serious work on metaphysics has only reawakened in the last generation or two. Many useful tools—supervenience, grounding, constitution, etc.—have only been around for a few years. And others, such as anchoring, are new. Even the metaphysical tools we already possess have only been applied to the social world by a few people. Moreover, anthropocentrism about the social world is a natural default position. As Kant pointed out, certain illusions are "inevitable" because of our subjective experience. Even the view that the sun revolves around the earth comes naturally out of our perspective. Likewise, it is perhaps inevitable that, as soon as we theorize about the social world—as contrasted with the natural world—we take it to be "made" by people in one way or another. Some things are a product of nature, and those that are not, it seems, must be a product of ourselves.

In this book I have aimed to dispel this illusion, and to make at least a start on rebuilding social ontology. Our distorted metaphysics of the social world is not the only ailment of the social sciences. People are too complicated for the social sciences ever to be easy. But of all the failings of the social sciences, its metaphysics is one we can make rapid progress on. It is frustrating to realize that our models have been built on distorted foundations, likely wasting time and effort, and inflicting damage on policymaking. On the other hand, this is also reason to be encouraged and optimistic about the future. Our theories of the social world are in their infancy, and it will not be hard to advance them.

A great many areas are in need of further investigation. Anchoring, in particular, is open terrain. There is work to be done on models of frames and possibilities, and in connection with this, we also need to investigate whether there are different varieties of anchoring,[1] and more fundamentally how anchoring works in the first place. I have suggested that individualism about anchoring is just as flawed as individualism about grounding is. But here too, the best argument is a better theory, and that remains to be developed. Anchoring, moreover, is dynamic, with new social properties and kinds introduced over time, and grounding conditions for social facts changing over time. How do we build on the materials from one frame, add some facts, and thereby anchor a new frame? This question remains largely unaddressed.

I have also only scratched the surface of the grounding inquiry. Much of the initial work in this book has been to clear the terrain and introduce the grounding–anchoring framework. In Part Two, I worked through the grounding of certain facts about groups, but did not consider nongroups in any serious way. We still need more tools to address things like corporations and universities, which stand in constitutive relations to many kinds of material. And then there are other sorts of objects—financial instruments, legal entities, artifacts, and so on—that are grounded differently still.

---

[1] I discuss this question in Epstein 2014b.

The discussion of group agency, as well, leaves as many questions as answers. I have discussed some ways in which a group can realize a system of practical activity, and argued that group attitudes are determined by more than the attitudes of group members. But I have not given a full theory of agency even for groups of one given kind. More also needs to be said about how a functional system, like our system of practical activity, gets realized by groups and by kinds of groups. Addressing this will, in part, be tied to improving our theories of anchoring. In particular, we will need to understand how functional roles and environmental facts can partly anchor a set of frame principles.

Finally, there is an even more important area for future research: how to move from an improved ontology to improved methods in social science. We can dramatically enhance our understanding of social entities. But we still need to see how we can use that to build better models.

Even when we recognize a flaw in the foundations of a model, it is always tempting just to add a patch. We saw this happen with Virchow. He observed that the body is composed by much more than cells, but then just added an epicycle: he expanded cells into "cell-territories." A patch like this may help temporarily. But when we move from a mere patch to a better overall ontology, our models see the benefits. No serious contemporary model of the body would be built on "cell-territories."

Getting the ontology right, rather than just patching it, is even more important in the social sciences. In the natural sciences, a distorted ontology inflicts damage. But the damage is worse in social science. The reason is that the natural sciences have a safety net, which the social sciences lack. In most models, we do not just represent the objects we are interested in targeting, that is, the objects that we are trying to understand or explain or predict. We also represent a variety of other objects that we take to be causally interacting with the targets. This gives us a certain amount of leeway in getting the ontology of our targets wrong. Even if we are sloppy about the ontology of our target objects, sometimes the grounds for facts about them get captured in the model anyway.

This sometimes rescues us in the natural sciences, because natural objects are usually confined to a particular region of space and time. Even though the body is not composed just of cells, for instance, it is not necessarily devastating to model it as though it were. A good model of cells will also include many things that the cells causally interact with. Therefore, the model as a whole is likely to capture a decent part of the actual ontology of the body, even though it does so in the wrong part of the model. This safety net is not foolproof, but it helps.

It does not work, however, in the social sciences. For the objects of the social sciences, we cannot assume that their constitutions are nearly so well behaved.

And as we have discussed, facts about social objects may be disconnected from facts about their constitutions. This means we cannot start with a distorted ontology and rely on causally related objects to fill the cracks.

In imitating the natural sciences, social modelers fail to recognize the differences between the objects of the natural sciences and those of the social sciences. Applying a more careful ontology to modeling in the social sciences means reworking the models. And potentially, it will lead us to whole classes of models that have been overlooked—ones that do not start with individuals, but with more heterogeneous grounds.[2]

There is much work to be done. Still, we should not lose sight of the simple point that motivates and unites it. It is tempting to see the social world either as a fabrication of our minds, or else as emerging from our attitudes and practices. These alternatives lose their appeal, as we develop and apply better tools. People do, of course, play a critical role in making the social world. Facts about people figure into anchoring frame principles and into grounding social facts. But to think the social world revolves around us—around our languages, our minds, our bodies, our practices—is old-school narcissism.

---

[2] I explore a few applications in Epstein 2008a, 2011, and Epstein and Forber 2012, but mostly this is an area for future work.

ACKNOWLEDGMENTS

This book has been in the works for a long time, and I am grateful to many people for their help and support. The Tufts Philosophy Department has been an ideal environment for research. I am grateful to Jody Azzouni, David Denby, Daniel Dennett, Erin Kelly, Dilip Ninan, and George Smith for detailed comments on parts of the manuscript. I am also grateful to Nancy Bauer, Avner Baz, Patrick Forber, Ray Jackendoff, Lionel McPherson, Christiana Olfert, Susan Russinoff, and Steven White for their help and friendship, and to Caleb Davis and Anne Belinsky for all their support.

I also want to express my appreciation to the remarkable graduate students and undergraduates at Tufts, especially those in my metaphysics and philosophy of social science classes. Thanks especially to Samia Hesni, David Laprade, Clare Saunders, Steven Norris, Colin Reeves, Alex Dietz, Oliver Traldi, and Jonathan Wright. Some of the initial ideas about individualism were sparked in a seminar I taught at Virginia Tech on philosophy of economics. I appreciate the contributions of the students and my former colleagues there.

I am grateful to the organizers of the Philosophy of Social Sciences Roundtable, the Collective Intentionality conferences, and other conferences, for giving me the opportunity to present and refine this material, and for their comments on my work, especially Finn Collin, Mattia Gallotti, Aki Lehtinen, Kirk Ludwig, John Michael, Mark Risjord, Paul Roth, Hans Bernhard Schmid, Raimo Tuomela, Stephen Turner, Alison Wylie, and Julie Zahle.

Many people have contributed their expertise in conversations and emails. I am particularly grateful to Lynne Baker, Karen Bennett, Johan van Benthem, Michael Bratman, Geoff Brennan, Margaret Gilbert, Kit Fine, Paul Griffiths, Davide Grossi, Till Grüne-Yanoff, Francesco Guala, Sally Haslanger, Frank Hindriks, Kevin Hoover, Harold Kincaid, Hartmut Kliemt, Christian List, Emiliano Lorini, Ruth Millikan, Laurie Paul, Sarah Paul, Philip Pettit, Gideon

Rosen, Don Ross, Abe Roth, David Schweikard, John Searle, Robert Stalnaker, Robert Sugden, Amie Thomasson, and Luca Tummolini.

Peter Ohlin and Emily Sacharin at Oxford University Press have made the publication process easy from the beginning, and I appreciate all their help. I am also very grateful to Paul Humphreys for his support, and to Constance Hale for her editorial flair.

Most of all, I'd like to thank my family, on whom I've relied beyond measure—Mom and Dad, Jonathan, Liu, Lea, Micah, and Jeremy; Barbara Rosenthal, Marty Birnbaum, Jack and Liz Rosenthal; and my friends, especially Bill MacCartney, Simon May, David Hornik, Mark Johnson, John Seybold, Cheryl Hemingway, Bill Fitzpatrick, Jerome Copulsky, and Mark Applebaum. And, of course, Sarah, my love.

BIBLIOGRAPHY

Agassi, J. 1975. "Institutional Individualism." *British Journal of Sociology* 26 (2):144–55.
Alexander, Jeffrey C., Bernhard Giesen, Richard Münch, and Neil J. Smelser, eds. 1987. *The Micro-Macro Link*. Berkeley: University of California Press.
Archer, Margaret S. 2003. *Structure, Agency and the Internal Conversation*. Cambridge: Cambridge University Press.
Ariew, André, Robert Cummins, and Mark Perlman. 2002. *Functions: New Essays in the Philosophy of Psychology and Biology*. Oxford: Oxford University Press.
Arrow, Kenneth J., and G. Debreu. 1954. "The Existence of an Equilibrium for a Competitive Economy." *Econometrica* 22 (3):265–90.
Audi, Paul. 2011. "Grounding: Toward a Theory of the in-Virtue-of Relation." *Journal of Philosophy* 109 (12):685–711.
Axtell, Robert. (2006) 2014. "Agent-Based Computing in Economics." Institute for New Economic Thinking at the Oxford Martin School, Oxford, January 21, 2014.
Baker, Lynne Rudder. 1997. "Why Constitution Is Not Identity." *Journal of Philosophy* 94 (12):599–621.
———. 1999. "Unity without Identity: A New Look at Material Constitution." In *New Directions in Philosophy: Midwest Studies in Philosophy*, edited by Peter A. French and Howard K. Wettstein, 144–65. Malden: Blackwell.
———. 2000. *Persons and Bodies: A Constitution View*. Cambridge: Cambridge University Press.
Banerjee, Abhijit V., Esther Duflo, Rachel Glennerster, and Cynthia Kinnan. (2010) 2013. "The Miracle of Microfinance? Evidence from a Randomized Evaluation." Working Paper No. 13–09.
Barro, Robert. 1973. "The Control of Politicians: An Economic Model." *Public Choice* 14:19–42.
Becker, Gary S., and George J. Stigler. 1974. "Law Enforcement, Malfeasance, and Compensation of Enforcers." *Journal of Legal Studies* 3 (1):1–18.
Beed, Clive, and Owen Kane. 1991. "What Is the Critique of the Mathematization of Economics?" *Kyklos* 44 (4):581–612.
Bender, Paul. 1962. "The Retroactive Effect of an Overruling Constitutional Decision: Mapp V. Ohio." *University of Pennsylvania Law Review* 110 (5):650–83.
Bennett, Karen. 2004a. "Global Supervenience and Dependence." *Philosophy and Phenomenological Research* 68 (3):510–29.
———. 2004b. "Spatio-Temporal Coincidence and the Grounding Problem." *Philosophical Studies* 118 (3):339–71.
———. 2009. "Composition, Colocation, and Metaontology." In *Metametaphysics: New Essays on the Foundations of Ontology*, edited by David Chalmers, David Manley, and Ryan Wasserman, 38–76. Oxford: Oxford University Press.

Bennett, Karen. 2011. "Construction Area: No Hard Hat Required." *Philosophical Studies* 154 (1):79–104.
Bernanke, Ben. (2004) 2012. "The Great Moderation." In *The Taylor Rule and the Transformation of Monetary Policy*, edited by Evan Koenig, Robert Leeson, and George Kahn, 145–62. Stanford: Hoover Institution Press.
Besley, Timothy. 2006. *Principled Agents? The Political Economy of Good Government*. Oxford: Oxford University Press.
Besley, Timothy, and Michael Smart. 2002. "Does Tax Competition Raise Voter Welfare?" *Discussion Paper Series Centre for Economic Policy Research* No. 3131.
Bhargava, Rajeev. 1992. *Individualism in Social Science*. Oxford: Clarendon.
Blackburn, Patrick, Johan van Benthem, and Frank Wolter. 2006. *Handbook of Modal Logic*. Vol. 3. Amsterdam: Elsevier Science.
Blanchard, Olivier J. 2008. "The State of Macro." *National Bureau of Economic Research* Working Paper No. 14259.
———. 2011. "Monetary Policy in the Wake of the Crisis." Paper presented at conference, "Macro and Growth Policies in the Wake of the Crisis," International Monetary Fund, Washington, DC, March 7–8, 2011. http://www.imf.org/external/np/seminars/eng/2011/res/pdf/OB2presentation.pdf.
Blanchard, Olivier J., Giovanni Dell'Ariccia, and Paolo Mauro. 2010. "Rethinking Macroeconomic Policy." *Journal of Money, Credit, and Banking* 42 (Supplement): 199–215.
Block, Ned. 1980. "Troubles with Functionalism." In *Readings in Philosophy of Psychology*, edited by Ned Block, 268–305. Cambridge, MA: Harvard University Press.
Bratman, Michael. 1987. *Intention, Plans, and Practical Reason*. Stanford: CSLI Publications.
———. 1993. "Shared Intention." *Ethics* 104 (1):97–113.
———. 1997. "I Intend That We J." In *Contemporary Action Theory*, Vol. 2: *Social Action*, edited by Raimo Tuomela and Ghita Holmstrom-Hintikka, 49–63. Dordrecht: Kluwer.
———. 2014. *Shared Agency*. Oxford: Oxford University Press.
Buller, David J. 1999. *Function, Selection, and Design*. Albany: SUNY Press.
Burke, Michael B. 1994. "Preserving the Principle of One Object to a Place." *Philosophy and Phenomenological Research* 54 (3):591–624.
Callender, Craig. 1999. "Reducing Thermodynamics to Statistical Mechanics: The Case of Entropy." *Journal of Philosophy* 96 (7):348–73.
Calvert, Randall. 1986. *Models of Imperfect Information in Politics*. Chur: Harwood Academic.
Canetti, Elias. 1962. *Crowds and Power*. New York: Viking.
Cassidy, John. 2010. "After the Blowup." *New Yorker*, January 11, 2010.
Colander, David. 1996. *Beyond Microfoundations: Post Walrasian Economics*. Cambridge: Cambridge University Press.
Coleman, James S. 1990. *Foundations of Social Theory*. Cambridge: Belknap Press.
Coleman, Jules, and Brian Leiter. 1996. "Legal Positivism." In *A Companion to Philosophy of Law and Legal Theory*, edited by Dennis Patterson, 241–50. Oxford: Blackwell.
Correia, Fabrice. 2005. *Existential Dependence and Cognate Notions*. Munich: Philosophia Verlag.
Currie, Gregory. 1984. "Individualism and Global Supervenience." *British Journal for the Philosophy of Science* 35 (4):345–58.
Dancy, Jonathan. 2004. *Ethics without Principles*. Oxford: Clarendon.
———. 2009. "Action, Content, and Inference." In *Wittgenstein and Analytic Philosophy: Essays for P. M. S. Hacker*, edited by John Hyman and Hans-Johann Glock, 278–98. Oxford: Oxford University Press.
Davidson, Donald. 1980. "Mental Events." *Readings in Philosophy of Psychology* 1:107–119.
de Tocqueville, Alexis. (1889) 2003. *Democracy in America*. Edited by Bruce Frohnen. Washington: Regenery.
Debreu, Gerard. 1991. "The Mathematization of Economic Theory." *American Economic Review* 81 (1):1–7.
deRosset, Louis. 2010. "Getting Priority Straight." *Philosophical Studies* 149 (1):73–97.
Doepke, Frederick. 1982. "Spatially Coinciding Objects." *Ratio* 24:45–60.

Doepke, Frederick. 1996. *The Kinds of Things: A Theory of Personal Identity Based on Transcendental Argument*. Chicago: Open Court.
Dorr, Cian, and Gideon Rosen. 2002. "Composition as a Fiction." In *The Blackwell Guide to Metaphysics*, edited by Richard Gale, 151–74. Oxford: Blackwell.
Drysdale, John James. 1874. *The Protoplasmic Theory of Life*. London: Bailliere, Tindall, and Cox.
Duflo, Esther. 2006. "Field Experiments in Development Economics." *Econometric Society Monographs* 42:322–48.
Durkheim, Emile. (1895) 1982. *The Rules of Sociological Method*. London: Macmillan.
Dworkin, Ronald. 1978. *Taking Rights Seriously*. Cambridge: Harvard University Press.
———. 1986. *Law's Empire*. Cambridge: Belknap Press.
Easterbrook, Frank H. 1983. "Statutes' Domains." *University of Chicago Law Review* 50 (2):533.
———. 1994. "Text, History, and Structure in Statutory Interpretation." *Harvard Journal of Law and Public Policy* 17:61.
Effingham, Nikk. 2010. "The Metaphysics of Groups." *Philosophical Studies* 149 (2):251–67.
Einheuser, Iris. 2006. "Counterconventional Conditionals." *Philosophical Studies* 127 (3):459–82.
Epstein, Brian. 2008a. "When Local Models Fail." *Philosophy of the Social Sciences* 38 (1):3–24.
———. 2008b. "The Realpolitik of Reference." *Pacific Philosophical Quarterly* 89 (1): 1–20.
———. 2009. "Ontological Individualism Reconsidered." *Synthese* 166 (1):187–213.
———. 2011. "Agent-Based Models and the Fallacies of Individualism." In *Models, Simulations, and Representations*, edited by Paul Humphreys and Cyrille Imbert, 115–44. New York: Routledge.
———. 2014a. "What Is Individualism in Social Ontology? Ontological Individualism vs. Anchor Individualism." In *Rethinking the Individualism-Holism Debate: Essays in the Philosophy of Social Science*, edited by Finn Collin and Julie Zahle, 17–38. Dordrecht: Springer.
———. 2014b. "How Many Kinds of Glue Hold the Social World Together?" In *Perspectives on Social Ontology and Social Cognition*, edited by Mattia Gallotti and John Michael, 41–55. Dordrecht: Springer.
Epstein, Brian, and Patrick Forber. 2012. "The Perils of Tweaking: How to Use Macrodata to Set Parameters in Complex Simulation Models." *Synthese* 190 (2):203–218.
Epstein, Joshua M. 2005. "Remarks on the Foundations of Agent-Based Generative Social Science." *Handbooks in Economics* 2:1585–604.
Farmer, J. Doyne, and Duncan Foley. 2009. "The Economy Needs Agent-Based Modelling." *Nature* 460 (7256):685–6.
Fearon, James. 1999. "Electoral Accountability and the Control of Politicians: Selecting Good Types Versus Sanctioning Poor Performance." In *Democracy, Accountability and Representation*, edited by Bernard Manin, Adam Przeworski, and Susan Stokes, 55–97. Cambridge: Cambridge University Press.
Ferejohn, John. 1986. "Incumbent Performance and Electoral Control." *Public Choice* 50 (1–3):5–25.
Fine, Kit. 2003. "The Non-Identity of a Material Thing and Its Matter." *Mind* 112 (446):195–234.
———. 2012. "Guide to Ground." In *Metaphysical Grounding: Understanding the Structure of Reality*, edited by Fabrice Correia and Benjamin Schnieder, 37–80. Cambridge: Cambridge University Press.
Fletcher, John. 1837. *Rudiments of Physiology: In Three Parts*. Edinburgh: J. Carfrae & Son.
Fodor, Jerry A. 1974. "Special Sciences (Or: The Disunity of Science as a Working Hypothesis)." *Synthese* 28 (2):97–115.
Forbes, G. 1987. "Is There a Problem of Persistence?" *Proceedings of the Aristotelian Society* 61 (Supplement):137–55.
Forel, Auguste. 1928. *The Social World of the Ants Compared with That of Man*. London: G. P. Putnam's Sons.
Garfinkel, Alan. 1981. *Forms of Explanation*. New Haven: Yale University Press.
Garfinkel, Harold. 1967. *Studies in Ethnomethodology*. Englewood Cliffs: Prentice-Hall.

Geertz, Clifford. 1973. *The Interpretation of Cultures: Selected Essays.* New York: Basic Books.
Gellner, Ernest. 1973. *Cause and Meaning in the Social Sciences.* London: Routledge & Kegan Paul.
Gibbard, Allan. 1975. "Contingent Identity." *Journal of Philosophical Logic* 4 (2):187–221.
Giddens, Anthony. 1984. *The Constitution of Society: Outline of the Theory of Structuration.* Cambridge: Polity.
Gilbert, Margaret. 1989. *On Social Facts.* Princeton: Princeton University Press.
———. 1990. "Walking Together: A Paradigmatic Social Phenomenon." *Midwest Studies in Philosophy* 15:1–14.
———. 1996. *Living Together: Rationality, Sociality, and Obligation.* Lanham: Rowman & Littlefield.
———. 2014. *Joint Commitment.* Oxford: Oxford University Press.
Goethe, J. W. (1816) 1962. *Italian Journey.* Translated by W. H. Auden and Elizabeth Mayer. London: Penguin.
Goldstein, Leon J. 1958. "The Two Theses of Methodological Individualism." *British Journal for the Philosophy of Science* 9 (33):1–11.
Green, Leslie. 1996. "The Concept of Law Revisited." *Michigan Law Review* 94 (6):1687–717.
Greenawalt, Kent. 1986. "The Rule of Recognition and the Constitution." *Michigan Law Review* 85:621–71.
Greenberg, Mark. 2004. "How Facts Make Law." *Legal Theory* 10 (3):158–98.
Grossi, Davide. 2007. "Designing Invisible Handcuffs: Formal Investigations in Institutions and Organizations for Multi-Agent Systems." *SIKS Dissertation Series* No. 16.
Grossi, Davide, John-Jules Ch. Meyer, and Frank Dignum. 2005. "Modal Logic Investigations in the Semantics of Counts-As." *Proceedings of the 10th International Conference on Artificial Intelligence and the Law,* 1–9.
———. 2006. "Classificatory Aspects of Counts-As: An Analysis in Modal Logic." *Journal of Logic and Computation* 16 (5):613–43.
———. 2008. "The Many Faces of Counts-As: A Formal Analysis of Constitutive Rules." *Journal of Applied Logic* 6 (2):192–217.
Guala, Francesco. 2007. "The Philosophy of Social Science: Metaphysical and Empirical." *Philosophy Compass* 2 (6):954–80.
Guala, Francesco, and Daniel Steel, eds. 2011. *The Philosophy of Social Science Reader.* London: Routledge.
Hale, Bob. 1987. *Abstract Objects.* Oxford: Blackwell.
Hamilton, William D. 1971. "Geometry for the Selfish Herd." *Journal of Theoretical Biology* 31 (2):295–311.
Hart, H. L. A. 1961. *The Concept of Law.* Oxford: Clarendon.
Haslanger, Sally. 1995. "Ontology and Social Construction." *Philosophical Topics* 23 (2):95–125.
Hawley, Katherine. 2006. "Principles of Composition and Criteria of Identity." *Australasian Journal of Philosophy* 84 (4):481–93.
Hawthorne, John. 2006. "Plenitude, Convention, and Ontology." In *Metaphysical Essays,* 53–69. Oxford: Oxford University Press.
Heims, Steve Joshua. 1993. *Constructing a Social Science for Postwar America: The Cybernetics Group, 1946–1953.* Cambridge: MIT Press.
Heller, Mark. 1990. *The Ontology of Physical Objects.* Cambridge: Cambridge University Press.
Hindriks, Frank. 2008. "The Status Account of Corporate Agents." In *Concepts of Sharedness: Essays on Collective Intentionality,* edited by Hans Bernd Schmid, K. Schulte-Ostermann, and N. Psarros, 119–44. Frankfurt: Ontos.
———. 2009. "Constitutive Rules, Language, and Ontology." *Erkenntnis* 71 (2):253–75.
———. 2012. "But Where Is the University?" *Dialectica* 66 (1):93–113.
———. 2013. "The Location Problem in Social Ontology." *Synthese* 190 (3):413–37.
Hirsch, Eli. 2002. "Against Revisionary Ontology." *Philosophical Topics* 30 (1):103–127.
Hooker, C. 1981. "Towards a General Theory of Reduction." *Dialogue* 20:496–529.
Hoover, Kevin. 1995. "Is Macroeconomics for Real?" *The Monist* 78 (3):235–57.
———. 2001a. *Causality in Macroeconomics.* Cambridge: Cambridge University Press.

Hoover, Kevin. 2001b. *The Methodology of Empirical Macroeconomics*. Cambridge: Cambridge University Press.
———. 2009. "Microfoundations and the Ontology of Macroeconomics." In *Oxford Handbook of Philosophy of Economics*, edited by Harold Kincaid and Don Ross, 386–409. Oxford: Oxford University Press.
Howe, Jeff. 2008. *Crowdsourcing: Why the Power of the Crowd Is Driving the Future of Business*. New York: Random House.
Hume, David. (1740) 1978. *A Treatise of Human Nature*. Edited by L. A. Selby-Bigge and P. H. Nidditch. 2nd ed. New York: Oxford University Press.
———. (1777) 1975. *Enquiries Concerning Human Understanding and Concerning the Principles of Morals*. edited by L. A. Selby-Bigge. Oxford: Clarendon.
Jarvie, I.C. 1998. "Situational Logic and Its Reception." *Philosophy of the Social Sciences* 28:365–80.
Johnston, Mark. 1992. "Constitution Is Not Identity." *Mind* 101:89–105.
Keynes, John Maynard. 1936. *The General Theory of Employment, Interest and Money*. New York: Harcourt, Brace.
Kim, Jaegwon. 1982. "Psychophysical Supervenience." *Philosophical Studies* 60 (2):51–70.
———. 1984. "Concepts of Supervenience." *Philosophy and Phenomenological Research* 45:153–76.
———. 1987. "'Strong' and 'Global' Supervenience Revisited." *Philosophy and Phenomenological Research* 48:315–26.
———. 1989. "The Myth of Nonreductive Materialism." *Proceedings and Addresses of the American Philosophical Association* 63:31–47.
Kincaid, Harold. 1986. "Reduction, Explanation and Individualism." In *Readings in the Philosophy of Social Science*, edited by M. Martin and L. C. McIntyre, 497–515. Cambridge: MIT Press.
———. 1997. *Individualism and the Unity of Science*. Lanham: Rowman & Littlefield.
———. 1998. "Supervenience." In *The Handbook of Economic Methodology*, edited by John B. Davis, D. Wade Hands, and Uskali Mäki, 487–8. Cheltenham: Edward Elgar.
Kirman, Alan. 2010. "The Economic Crisis Is a Crisis for Economic Theory." *CESifo Economic Studies* 56 (4):498–535.
Klein, L. R., and A. S. Goldberger. 1955. *An Econometric Model of the United States, 1929–1952*. Amsterdam: North-Holland.
Koslicki, Kathrin. 2004. "Constitution and Similarity." *Philosophical Studies* 117:327–64.
———. 2012. "Varieties of Ontological Dependence." In *Metaphysical Grounding: Understanding the Structure of Reality*, edited by Fabrice Correia and Benjamin Schnieder, 186–213. Cambridge: Cambridge University Press.
Kratzer, Angelika. 2012. *Modals and Conditionals: New and Revised Perspectives*. Oxford: Oxford University Press.
Kripke, Saul. (1972) 1980. *Naming and Necessity*. Cambridge: Harvard University Press.
Krugman, Paul. 2009. "How Did Economists Get It So Wrong?" *New York Times Magazine*, September 2, 2009.
Kuhn, Thomas S. 1962. *The Structure of Scientific Revolutions*. Chicago: University of Chicago Press.
Kydland, Finn E., and Edward C. Prescott. 1982. "Time to Build and Aggregate Fluctuations." *Econometrica* 50 (6):1345–70.
Langton, Rae, and David K. Lewis. 1998. "Defining 'Intrinsic'." *Philosophy and Phenomenological Research* 58:333–45.
Lattimore, Richmond. 1951. *The Iliad of Homer*. Chicago: University of Chicago Press.
Le Bon, Gustave. 1895. *The Crowd: A Study of the Popular Mind*. New York: Macmillan.
Leiter, Brian. 2003. "Beyond the Hart/Dworkin Debate: The Methodology Problem in Jurisprudence." *American Journal of Jurisprudence* 48:17–51.
Leuenberger, Stephan. 2014. "Grounding and Necessity." *Inquiry* 57 (2):151–74.
Lewis, David K. 1969. *Convention: A Philosophical Study*. Cambridge, MA: Harvard University Press.
———. 1973. *Counterfactuals*. Oxford: Blackwell.

Lewis, David K. 1976. "Survival and Identity." In *The Identities of Persons*, edited by Amelie Oksenberg Rorty, 17–40. Berkeley: University of California Press.
———. 1983. "Extrinsic Properties." *Philosophical Studies* 44:197–200.
———. 1986. *On the Plurality of Worlds*. Cambridge: Cambridge University Press.
List, Christian, and Philip Pettit. 2006. "Group Agency and Supervenience." *Southern Journal of Philosophy* 44:85–105.
———. 2011. *Group Agency: The Possibility, Design, and Status of Corporate Agents*. Oxford: Oxford University Press.
List, Christian, and Kai Spiekermann. 2013. "Methodological Individualism and Holism in Political Science: A Reconciliation." *American Political Science Review* 107:629–43.
Little, Daniel. 1991. *Varieties of Social Explanation*. Boulder: Westview.
Lo, Andrew. 2012. "Reading about the Financial Crisis: A Twenty-One-Book Review." *Journal of Economic Literature* 50 (1):151–78.
Lo, Andrew, and Mark Mueller. 2010. "Warning: Physics Envy May Be Hazardous to Your Wealth!" *Journal of Investment Management* 8 (2):13–63.
Lowe, E. J. 1989. *Kinds of Being: A Study of Individuation, Identity, and the Logic of Sortal Terms*. Oxford: Blackwell.
———. 1991. "One-Level Criteria of Identity and Two-Level Criteria of Identity." *Analysis* 51:192–4.
———. 1995. "Coinciding Objects: In Defence of the 'Standard Account'." *Analysis* 55 (3):171–8.
———. 1997. "Objects and Criteria of Identity." In *A Companion to the Philosophy of Language*, edited by Bob Hale and Crispin Wright, 613–33. Oxford: Blackwell.
———. 1998. *The Possibility of Metaphysics: Substance, Identity, and Time*. Oxford: Oxford University Press.
———. 2009. *More Kinds of Being*. Oxford: Wiley-Blackwell.
Lowe, E. J., and Harold W. Noonan. 1988. "Substance, Identity and Time." *Proceedings of the Aristotelian Society* 62 (Supplement):61–100.
Lucas, Robert E. 1972. "Expectations and the Neutrality of Money." *Journal of Economic Theory* 4 (2):103–124.
———. 1977. "Understanding Business Cycles." *Carnegie-Rochester Conference Series on Public Policy* 5:7–29.
Lukes, Steven. 1968. "Methodological Individualism Reconsidered." *British Journal of Sociology* 19:119–29.
Macdonald, Graham, and Philip Pettit. 1981. *Semantics and Social Science*. London: Routledge & Kegan Paul.
Mandelbaum, Maurice. 1955. "Societal Facts." *British Journal of Sociology* 6:305–317.
Mandeville, Bernard. (1714) 1934. *The Fable of the Bees; or, Private Vices, Public Benefits*. London: Wishart.
Manning, John F. 2005. "Textualism and Legislative Intent." *Virginia Law Review* 91: 419–50.
Marmor, Andrei. 2010. *Philosophy of Law*. Princeton: Princeton University Press.
———. 2012. "The Nature of Law: An Introduction." In *The Routledge Companion to the Philosophy of Law*, edited by Andrei Marmor, 3–15. New York: Routledge.
Marx, Karl. 1867. *Capital*. Vol. 1. Harmondsworth: Penguin.
Marx, Karl, and Friedrich Engels. (1975) 1998. *Collected Works*. Vol. 37. London: Lawrence & Wishart.
Marx, Maarten, and Yde Venema. 1997. *Multi-Dimensional Modal Logic*. Dordrecht: Kluwer Academic.
May, Larry. 1992. "On Social Facts by Margaret Gilbert." *Ethics* 102 (4):853–6.
Mayr, Ernst. 1982. *The Growth of Biological Thought: Diversity, Evolution, and Inheritance*. Cambridge: Belknap Press.
McLaughlin, Brian P. 1984. "Perception, Causation, and Supervenience." *Midwest Studies in Philosophy* 9 (1):569–92.
———. 1995. "Varieties of Supervenience." In *Supervenience: New Essays*, edited by E. Savellos and U. Yalcin, 16–59. Cambridge: Cambridge University Press.

McLaughlin, Brian, and Karen Bennett. 2005. "Supervenience." Stanford Encyclopedia of Philosophy. Accessed August 10, 2014. http://plato.stanford.edu/archives/win2011/entries/supervenience/.
Mellor, D. H. 1982. "The Reduction of Society." *Philosophy* 57:51–75.
Merricks, Trenton. 2001. *Objects and Persons*. Oxford: Oxford University Press.
Mill, John Stuart. (1843–72) 1974. *A System of Logic. Ratiocinative and Inductive*. Vols. 7–8 of *Collected Works*. Toronto: University of Toronto Press.
Mills, C. Wright. 1959. *The Sociological Imagination*. New York: Oxford University Press.
Nagel, Ernest. 1961. *The Structure of Science: Problems in the Logic of Scientific Explanation*. London: Routledge & Kegan Paul.
Nash, John F. 1950. "The Bargaining Problem." *Econometrica* 18 (2):155–62.
Neale, Stephen. 2001. *Facing Facts*. New York: Oxford University Press.
Nickles, Thomas. 1973. "Two Concepts of Intertheoretic Reduction." *Journal of Philosophy* 70 (7):181–201.
Noonan, Harold. 1993. "Constitution Is Identity." *Mind* 102:133–46.
North, Michael, and Charles Macal. 2007. *Managing Business Complexity: Discovering Strategic Solutions with Agent-Based Modeling and Simulation*. New York: Oxford University Press.
Nøttestad, Leif, and Bjørn Erik Axelsen. 1999. "Herring Schooling Manoeuvres in Response to Killer Whale Attacks." *Canadian Journal of Zoology* 77 (10):1540–46.
Oderberg, D. S. 1996. "Coincidence under a Sortal." *Philosophical Review* 105:145–71.
Oz, Amos. 2013. *Between Friends*. Translated by Sondra Silverston. New York: Houghton Mifflin Harcourt.
Parsons, Talcott. 1951. *The Social System*. Glencoe: Free Press.
———. 1954. *Essays in Sociological Theory*. Glencoe: Free Press.
Paul, L. A. 2002. "Logical Parts." *Noûs* 36 (4):578–96.
———. 2010. "The Puzzles of Material Constitution." *Philosophy Compass* 5 (7):579–90.
Paul, Sarah K. 2013. "The Conclusion of Practical Reasoning: The Shadow between Idea and Act." *Canadian Journal of Philosophy* 43 (3):287–302.
Peacocke, Christopher. 2005. "Joint Attention: Its Nature, Reflexivity, and Relation to Common Knowledge." In *Joint Attention: Communication and Other Minds*, edited by Naomi Eilan, Christoph Hoerl, Teresa McCormack, and Johannes Roessler, 298–324. New York: Oxford University Press.
Pettit, Philip. 1993. *The Common Mind*. New York: Oxford University Press.
———. 2003. "Groups with Minds of Their Own." In *Socializing Metaphysics*, edited by Frederick Schmitt, 167–94. Lanham: Rowman & Littlefield.
Pitcher, T. J., and J. K. Parrish. 1993. "Functions of Shoaling Behavior in Teleosts." In *The Behaviour of Teleost Fishes*, edited by T. J. Pitcher, 364–439. London: Croom Helm.
Plato. 1969. *Republic*. Translated by Paul Shorey. Edited by Paul Shorey. Vol. 2. Cambridge: Harvard University Press.
Popper, Karl. 1945. *The Open Society and Its Enemies*. London: Routledge & Kegan Paul.
———. (1957) 2002. *The Poverty of Historicism*. London Routledge.
Portner, Paul. 2009. *Modality*. Oxford: Oxford University Press.
Prakash, S. B., and J. C. Yoo. 2003. "The Origins of Judicial Review." *University of Chicago Law Review* 69:887–982.
Pufendorf, Samuel. (1673) 2007. *On the Duty of Man and Citizen According to Natural Law*. Cambridge: Cambridge University Press.
Putnam, Hilary. 1967. "Psychological Predicates." In *Art, Mind, and Religion*, edited by W. Capitan and D. D. Merrill, 37–48. Pittsburgh: University of Pittsburgh Press.
———. 1969. "On Properties." In *Essays in Honor of Carl G. Hempel*, edited by Nicholas Rescher, 235–54. Dordrecht: D. Reidel.
Quine, W. V. 1960. *Word and Object*. Cambridge: MIT Press.
———. 1981. *Theories and Things*. Cambridge, MA: Harvard University Press.
Ranke, Leopold von. (1836) 1981. *The Secret of World History: Selected Writings on the Art and Science of History*. Translated by Roger Wines. New York: Fordham University Press.
Raz, Joseph. 1975. *Practical Reason and Norms*. Princeton: Princeton University Press.

Rea, Michael C. 2000. "Constitution and Kind Membership." *Philosophical Studies* 97 (2):169–93.
Richard, Mark. 1990. *Propositional Attitudes*. Cambridge: Cambridge University Press.
Ritchie, Katherine. 2013. "What Are Groups?" *Philosophical Studies* 166:257–72.
Robins, Michael. 2002. "Joint Commitment and Circularity." In *Social Facts and Collective Intentionality*, edited by Georg Meggle, 299–321. Frankfurt: Hansel-Hohenhausen.
Robinson, Denis. 1982. "Re-Identifying Matter." *Philosophical Review* 81:317–42.
Romer, Paul. 2011. "The Dynamics of Rules." Paper presented at conference, "Macro and Growth Policies in the Wake of the Crisis," International Monetary Fund, Washington, DC, March 7–8, 2011. http://www.imf.org/external/np/seminars/eng/2011/res/pdf/PRpresentation.pdf.
Rose-Ackerman, Susan. 1978. *Corruption: A Study in Political Economy*. New York: Academic Press.
Rosen, Gideon. 2010. "Metaphysical Dependence: Grounding and Reduction." In *Modality: Metaphysics, Logic, and Epistemology*, edited by Bob Hale and Aviv Hoffman, 109–136. Oxford: Oxford University Press.
Ross, Don. 2008. "Two Styles of Neuroeconomics." *Economics and Philosophy* 24 (3):473–83.
Sawyer, R. Keith. 2002. "Nonreductive Individualism: Part I." *Philosophy of the Social Sciences* 32 (4):537–59.
———. 2005. *Social Emergence*. Cambridge: Cambridge University Press.
Schaffer, Jonathan. 2012. "Grounding, Transitivity, and Contrastivity." In *Metaphysical Grounding: Understanding the Structure of Reality*, edited by Fabrice Correia and Benjamin Schnieder, 122–38. Cambridge: Cambridge University Press.
Schaffner, Kenneth F. 1967. "Approaches to Reduction." *Philosophy of Science* 34:137–47.
———. 1993. "Theory Structure, Reduction, and Disciplinary Integration in Biology." *Biology and Philosophy* 8 (3):319–47.
Schmitt, Frederick. 2003. "Socializing Metaphysics: An Introduction." In *Socializing Metaphysics*, edited by Frederick Schmitt, 1–37. Lanham: Rowman & Littlefield.
Schnapp, Jeffrey Thompson, and Matthew Tiews. 2006. *Crowds*. Stanford: Stanford University Press.
Schwann, Theodor. (1839) 1847. *Microscopical Researches into the Accordance in the Structure and Growth of Animals and Plants*. Translated by Henry Smith. London: Sydenham Society.
Searle, John R. 1995. *The Construction of Social Reality*. New York: Free Press.
———. 2008. "Language and Social Ontology." *Theory and Society* 37 (5):443–59.
———. 2010. *Making the Social World: The Structure of Human Civilization*. Oxford: Oxford University Press.
Shagrir, Oren. 2002. "Global Supervenience, Coincident Entities, and Anti-Individualism." *Philosophical Studies* 109:171–96.
Shah, Nishi. 2008. "How Action Governs Intention." *Philosophers' Imprint* 8 (5):1–19.
Shapiro, Lawrence. 2000. "Multiple Realizations." *Journal of Philosophy* 97 (12):635–54.
Shapiro, Scott. 2007. "The 'Hart-Dworkin' Debate: A Short Guide for the Perplexed." In *Ronald Dworkin*, edited by A. Ripstein, 22–55. Cambridge: Cambridge University Press.
———. 2009. "What Is the Rule of Recognition (and Does It Exist)?" In *The Rule of Recognition and the U.S. Constitution*, edited by Matthew Adler and Kenneth Himma, 235–68. New York: Oxford University Press.
———. 2011. *Legality*. Cambridge: Belknap Press.
Sheehy, Paul. 2002. "On Plural Subject Theory." *Journal of Social Philosophy* 32:377–94.
Shoemaker, Sydney. 1999. "Self, Body, and Coincidence." *Proceedings of the Aristotelian Society, Supplementary Volumes* 73:287–306.
———. 2003. "Realization, Micro-Realization, and Coincidence." *Philosophy and Phenomenological Research* 67 (1):1–23.
Sider, Theodore. 1999. "Global Supervenience and Identity across Times and Worlds." *Philosophy and Phenomenological Research* 59:913–37.
———. 2001. *Four-Dimensionalism: An Ontology of Persistence and Time*. Oxford: Oxford University Press.

Sider, Theodore. 2002. "The Ersatz Pluriverse." *Journal of Philosophy* 99 (6):279–315.

———. 2006. "Yet Another Paper on the Supervenience Argument against Coincident Entities." *Philosophy and Phenomenological Research* 77 (3):613–24.

———. 2009. "Ontological Realism." In *Metametaphysics: New Essays on the Foundations of Ontology*, edited by David John Chalmers, David Manley, and Ryan Wasserman, 384–423. Oxford: Oxford University Press.

Simons, Peter. 1987. *Parts: A Study in Ontology*. Oxford: Oxford University Press.

Skiles, Alexander. 2014. "Against Grounding Necessitarianism." *Erkenntnis* 79 (9).

Sleigh, Charlotte. 2007. *Six Legs Better: A Cultural History of Myrmecology*. Baltimore: Johns Hopkins University Press.

Smith, Adam. (1776) 2006. *An Inquiry into the Nature and Causes of the Wealth of Nations*. Cirencester: Echo Library.

Smith, Barry. 2003. "John Searle: From Speech Acts to Social Reality." In *John Searle*, edited by Barry Smith, 1–33. Cambridge: Cambridge University Press.

———. 2008. "Searle and De Soto: The New Ontology of the Social World." In *The Mystery of Capital and the Construction of Social Reality*, edited by Barry Smith, David Mark, and Isaac Ehrlich, 35–51. Chicago: Open Court.

Spencer, Herbert. 1895. *The Principles of Sociology*. Vol. 1. New York: Appleton.

Stalnaker, Robert. 1968. "A Theory of Conditionals." In *Studies in Logical Theory*, 98–112. American Philosophical Quarterly Monograph Series, 2. Oxford: Blackwell.

———. 1996. "Varieties of Supervenience." *Philosophical Perspectives* 10:221–41.

Strevens, Michael. 2007. "Review of Woodward, Making Things Happen." *Philosophy and Phenomenological Research* 74 (1):233–49.

Suppe, Frederick. 1972. "What's Wrong with the Received View on the Structure of Scientific Theories?" *Philosophy of Science* 39 (1):1–19.

———. 1977. *The Structure of Scientific Theories*. Urbana: University of Illinois Press.

Suppes, Patrick. 1967. "What Is a Scientific Theory?" In *Philosophy of Science Today*, edited by S. Morgenbesser, 55–67. New York: Basic Books.

Surowiecki, James. 2005. *The Wisdom of Crowds*. New York: Anchor.

Szwed, John. 2002. *So What: The Life of Miles Davis*. New York: Simon and Schuster.

Thalos, Mariam. 2013. *Without Hierarchy: The Scale Freedom of the Universe*. New York: Oxford University Press.

Thomson, Judith Jarvis. 1983. "Parthood and Identity across Time." *Journal of Philosophy* 80:201–220.

———. 1998. "The Statue and the Clay." *Noûs* 32:149–73.

Thoreau, Henry David. (1854) 1966. *Walden*. New York: W. W. Norton.

Tinbergen, Jan. 1956. *Economic Policy: Principles and Design*. Amsterdam: North-Holland.

Tollefsen, Deborah. 2002. "Organizations as True Believers." *Journal of Social Philosophy* 33 (3):395–411.

Tuomela, Raimo. 1989. "Collective Action, Supervenience, and Constitution." *Synthese* 80: 243–66.

———. 1995. *The Importance of Us: A Philosophical Study of Basic Social Notions*. Stanford: Stanford University Press.

———. 2002. *The Philosophy of Social Practices: A Collective Acceptance View*. Cambridge: Cambridge University Press.

———. 2007. *The Philosophy of Sociality: The Shared Point of View*. Oxford: Oxford University Press.

———. 2011. "Searle's New Construction of Social Reality." *Analysis* 71 (4):706–719.

———. 2013. *Social Ontology*. Oxford: Oxford University Press.

Tuomela, Raimo, and Kaarlo Miller. 1988. "We-Intentions." *Philosophical Studies* 53 (3): 367–89.

———. 2005. "We-Intentions Revisited." *Philosophical Studies* 125 (3):327–69.

Udehn, Lars. 2001. *Methodological Individualism: Background, History, and Meaning*. London: Routledge.

———. 2002. "The Changing Face of Methodological Individualism." *Annual Review of Sociology* 28:479–507.

Unger, Peter. 1979. "There Are No Ordinary Things." *Synthese* 41 (2):117–54.
Uzquiano, Gabriel. 2004. "The Supreme Court and the Supreme Court Justices: A Metaphysical Puzzle." *Noûs* 38 (1):135–53.
van Benthem, Johan. 1996. *Exploring Logical Dynamics*. Stanford: CSLI Publications.
van Cleve, James. 1990. "Supervenience and Closure." *Philosophical Studies* 58 (3):225–38.
van Fraassen, Bas C. 1972. *A Formal Approach to the Philosophy of Science*. Pittsburgh: University of Pittsburgh Press.
———. 1977. "The Pragmatics of Explanation." *American Philosophical Quarterly* 14 (2): 143–50.
———. 1987. "The Semantic Approach to Scientific Theories." In *The Process of Science*, edited by N. J. Nersessian, 105–124. Dordrecht: Nijhoff.
van Inwagen, Peter. 1987. "When Are Objects Parts?" *Philosophical Perspectives* 1:21–47.
———. 1990. *Material Beings*. Ithaca: Cornell University Press.
Virchow, Rudolf Ludwig Karl. 1860. *Cellular Pathology: As Based Upon Physiological and Pathological Histology*. Translated by Franklin Chance. London: John Churchill.
von Neumann, John, and Oskar Morgenstern. 1944. *Theory of Games and Economic Behavior*. Princeton: Princeton University Press.
Wallace, Walter. 1996. "A Definition of Social Phenomena for the Social Sciences." In *The Mark of the Social: Discovery or Invention*, edited by John Greenwood, 37–58. Lanham: Rowman & Littlefield.
Wasserman, Ryan. 2002. "The Standard Objection to the Standard Account." *Philosophical Studies* 111 (3):197–216.
———. 2004. "The Constitution Question." *Noûs* 38 (4):693–710.
Watkins, J. W. N. 1953. "The Principle of Methodological Individualism." *British Journal for the Philosophy of Science* 3:186–9.
———. 1955. "Methodological Individualism: A Reply." *Philosophy of Science* 22:58–62.
———. 1959. "The Two Theses of Methodological Individualism." *British Journal for the Philosophy of Science* 9:319–20.
Weber, Marcel. 2004. *Philosophy of Experimental Biology*. Cambridge: Cambridge University Press.
Weber, Max. 1978. *Economy and Society*. Translated by H. H. Gerth and C. Wright Mills. Edited by Guenther Roth and Claus Wittich. Berkeley: University of California Press.
Wiggins, David. 1968. "On Being in the Same Place at the Same Time." *Philosophical Review* 77 (1):90–95.
———. 2001. *Sameness and Substance Renewed*. Cambridge: Cambridge University Press.
Williamson, Timothy. 1991. "Fregean Directions." *Analysis* 51:194–5.
———. 2013. *Identity and Discrimination*. Oxford: Wiley-Blackwell.
Wimsatt, William. 1976. "Reductionism, Levels of Organization, and the Mind-Body Problem." In *Consciousness and the Brain*, edited by Gordon G. Globus, Grover Maxwell, and Irwin Savodnik, 205–263. New York: Plenum.
———. 1994. "The Ontology of Complex Systems: Levels of Organization, Perspectives, and Causal Thickets." *Canadian Journal of Philosophy* 20 (Supplement):207–274.
Wisdom, J. O. 1970. "Situational Individualism and the Emergent Group Properties." In *Explanation in the Behavioural Sciences*, edited by R. Borger and F. Cioffi, 271–96. Cambridge: Cambridge University Press.
Witmer, D. Gene, William Butchard, and Kelly Trogdon. 2005. "Intrinsicality without Naturalness." *Philosophy and Phenomenological Research* 70 (2):326–50.
Wynne-Edwards. 1962. *Animal Dispersion in Relation to Social Behaviour*. Edinburgh: Oliver & Boyd.
Yablo, Stephen. 1987. "Identity, Essence, and Indiscernibility." *Journal of Philosophy* 84: 293–314.
———. 1999. "Intrinsicality." *Philosophical Topics* 26:479–505.
Zangwill, Nick. 2008. "Moral Dependence." *Oxford Studies in Metaethics* 3:109–127.

INDEX

actual fact inquiry, 77, 99–100
anchor individualism
  defined, 101, 103–6
  distinct from ontological
    individualism, 124–26
  is probably false, 105–6, 123, 130, 151, 162
    n8, 193, 218, 235 n16, 273, 277–78
  Hart's theory as, 95–97, 104
  Hume's theory as, 54–55, 104
  Searle's theory as, 103–4, 126
  weak version of, 104–5
anchoring, 80–87, 90–99, 103–5, 115–28, 157,
    167–68, 217–18, 236–39
  defined, 80–85
  functions in, 68, 219, 237–38, 240, 244,
    248–49, 262, 265, 278
  historical precursors, 53–54, 54 n9, 82, 105
  of intention and action, 201–2, 217–18,
    236–39, 248–49, 272
  of a law, 90–95, 162 n8
  often inexplicit, 153 n1, 193, 235 n16
  relation to grounding, 82, 86–87, 115–28
  of the rule of recognition, 95–96, 98–99
anchoring inquiry, 86–87, 100, 105–6, 127–28
  defined, 86–87
  in the law, 94–96, 98, 100
  less central to modeling, 128
ant colonies, 5, 7, 26–27, 136–37, 163, 167, 260
anthropocentrism, 6–8, 127, 129–30, 199,
    247, 260
  as natural illusion, 277

Bennett, K., 33, 69, 111, 142, 227
Bernanke, B., 2–3
Block, N., 214–15
boat diagram, 42–44
Bratman, M.
  modest sociality, 253–54

roles of intention, 201–2, 217–18,
  233, 237–38
theory of shared intention, 8, 200–3, 207–8,
  243, 259
bridge laws, 28
Bulger, W., 90–91, 93–97, 99
Burke, E., 4–5

Catholic Church, 233
causal explanations, 27–29, 31, 43–45,
  128, 278–79
causation, 26, 45 n16, 81
  distinct from grounding, 69, 72, 106, 165,
    184–85, 224, 230–33
  may be among anchors, 193
  may be among grounds, 190–91, 255
cell territories, 39–41, 45, 48, 278
cell theory, 37–40
China-body, 214–16, 217, 227
Coleman, J., 42–46
collective acceptance, 52–53, 74, 81–82,
  103, 117–18
  among grounding conditions, 117–18,
    123 n11
  Searle's theory of, 52–53, 55, 95, 103–4, 126
  as theory of anchoring, 53, 55, 81–82,
    103–4, 118, 119, 123, 264, 269–71
  in Tuomela, 121 n8, 255, 266–68
collective intention. *See* group intention
conjunctivism, 82, 115–27, 157 n5, 160,
  220, 241
  compatible with remainder of book, 127, 157
    n5, 160
  counterfactual not evidence for, 118–20
  defined, 115–16
  the fundamental flaw with, 123–24
  is a problem for individualist, 220, 241
connectability, 30–32

constitution, 10, 75 n1, 133, 138–49
   and coincidence, 139–42
   group-constitution, 145–46
   new analysis of, 146–49
   of a legislature, 230–33
   of a rock, 163–66
   of Supreme Court, 153–58
   of a team, 190–91, 204–5
   Thomson's theory of, 142–44
   usually non-intrinsic, 210, 216, 220, 260
constitutive rule
   clarified, 58, 74–77
   frame principle as generalization of, 76–80
   Searle's definition, 52–53
   structural flaws of, 120–23, 161, 264, 268
   X term of, 52, 58, 75–76, 126
convention, 53–55, 76, 104–5, 119
   anchors of, 54–55, 81, 95, 104
   Hume on, 53–55, 97
   illustrating generalized anchor individualism, 105
   Lewis on, 95, 97
   as species of frame principle, 95, 97
corporate agents, 142 n24, 265, 268–72
counterfactuals, 117–20
criteria of identity
   the basic idea, 170–72
   clarified, 172–74
   derivable from frame principles, 194–95
   do not give grounds, 181
   generalized, 174–80
   usefulness of, 169–70, 180–81
criteria of legal validity, 97–99
cross-identifying criteria, 178–80
crowd. *See* mob
dependence
   defined, 106–9
   frame principles expressing, 157, 160, 200 n12
   and ontological individualism, 109–10, 124–25
   relation to determination, 107–9
   supervenience as evidence for, 72, 111–12, 224
   various claims involving, 20, 41, 46, 155, 236 n1, 238

Descartes, R., 17–18
determination
   defined, 71, 106–9, 116 n1
   frame principles expressing, 156–57, 186, 191, 200 n12
   overlooked by Rosen, 107 n7
   relation to dependence, 72, 107–9
   various claims involving, 17–19, 21, 29, 34–35, 46, 59, 214

Doepke, F., 147–48
dualism, 17–18, 25, 50, 112
   anxiety about, 13–14, 20, 29, 35
   two strategies for avoiding, 22, 30, 32–34, 40, 50–51, 58–59, 126
Durkheim, E., 8, 197–98
Dworkin, R., 97–98

Easterbrook, F., 199–200, 202–4, 239
electoral control, 230–33, 245–47
emergentism, 5, 23–24
   different from supervenience failure, 166–67
   and individualism, 37, 126, 128
   as theory of social objects, 25–27, 35, 279
epicycles, 7, 48–49, 126, 197, 278
existence. *See under* grounding conditions
explanatory individualism, 21–22, 35
externalism, 48–49, 214, 236 n1

facts, 61–63, 66–69, 75, 102–3, 150–53
   and grounding, 69–70
   individualistic (*see* individualistic facts and properties)
   social (*see* social facts and properties)
   and supervenience, 112–14
Fine, K., 71, 140, 141
Fodor, J., 31–32
frame principles, 77–80, 103–5, 116
   anchoring of, 80–84, 101, 103–5, 127
   defined, 77–80
   examples worked through, 156–63, 186–93
   as generalization of constitutive rules, 74–80
   giving grounding conditions, 76–78, 115–16
   laws as, 88, 90–91, 94–97, 99
   in social integrate model, 252, 258
   some forms of, 106, 151–53, 156–57
   typically applying to all possibilities, 78, 116, 116 n1, 123–24, 156 n5, 157
frames, 77–80, 96–97, 103–4, 116
   defined, 78, 116
   multiple, 86, 103, 116, 119–20, 156 n5, 277
   nested, 96–97
   as universes of possible worlds, 78–79, 116
functional kinds
   in anchoring, 15, 68, 278
   and multiple realizability, 31–32, 248–49, 262–63, 272
   standard examples of, 15, 31–32, 214–15
   and status assignments, 51–52, 264–65
   in system of practical activity, 237–38, 240, 244, 248–49, 262

Garfinkel, H., 16
generics, 75

Giddens, A., 49
Gilbert, M., 8, 135–36, 139 n14, 200, 250–53, 255–57
Goethe, J.W., 40
grounding, 8, 69–72, 80–81, 106–12, 204–13
  full versus partial, 70–71, 85, 107
  necessitarian and contingency views, 71, 156 n5; irrelevance to conjunctivism, 116 n1; skepticism about necessitarianism, 71, 107 n6
  patterns of, 204–13
  relations analyzed in terms of: constitution, 142, 148–49; dependence, 72, 107–9; determination, 106–7; ontological individualism, 109–10
  and supervenience, 72, 110–12
grounding-anchoring model
  applied to law, 93–97
  and grounding problem, 227 n5
  for interpreting counterfactuals, 119–20
  introduced, 82–87
grounding conditions
  conjunctivism gets wrong, 123–24
  of constitution, 153–58, 164–66, 190–91, 231–33
  defined, 76–80
  of existence, 76, 147, 151, 158–61, 181, 184–93, 252, 255, 258–59
  of group action, 219–27, 231–33
  of group intention, 239–47
  often diachronic, 185–86, 252
  often heterogeneous, 157, 163, 167–68, 190, 193, 204–13, 217
  of various facts, 162–63, 204–13
grounding inquiry, 86–87, 99–100, 127–29
  defined, 86–87
  in the law, 99–100
  role in modeling, 100, 127–28
  as topic of Part Two, 129
group action
  intertwined with intention, 217–18, 237–39
  more than member, 219–34
  new perspective on, 234–35
group attitudes
  introduced, 197–99
  List and Pettit on, 213–16
group intention
  and action: intertwined with, 217–18, 237–39; need not perfectly align, 241–42
  Bratman's theory, 200–1, 207–9
  member intentions insufficient for, 239–47, 262–63
  naive theory, 199–200
  relation to individual intention, 200–2, 217–18, 237–38, 248–49
  roles of, 201–2, 217–18, 233

groups
  broad conception of, 133, 149, 251, 257–58, 262–63
  constitution of, 144–46, 149, 153–58, 190–91, 204–5, 230–33
  existence of, 147, 158–61, 184–93
  as focus of book's grounding inquiry, 9–10, 80, 129–31
  paradigms of, 22, 24–27, 47, 56–57, 126, 133–36
  mathematical models of, 136–38, 145
  ordinary object model of, 138–39, 144–46
  social integrate models of, 250–63
  status models of, 264–72
Guala, F., 37 n2, 50, 55

Hamilton, W.D., 25, 27–29
Hart, H.L.A., 88–90, 94–99, 101, 104
Haslanger, S., 8, 69
Hayek, F., 4
Hegel, G.W.F., 13–14
herring, 24–29
hierarchy, in groups, 219–24, 239–42, 255–57
high-level properties. *See under* properties
Hindriks, F., 37 n2, 265, 268–69, 272
hodgepodge objection, 199–200, 202
holism, 13–14, 19–20, 48
  no need to endorse, 37
House of Representatives, 221, 245–47
Hume, D., 53–55, 76, 81, 97, 104
identity, 140, 147, 169–81, 260–61
  properties, 113 (*see also* criteria of identity)
individualism
  anchor (*see under* anchor individualism)
  explanatory, 21–22, 35
  methodological, 18–21, 34–35
  nonreductive, 34
  ontological (*see under* ontological individualism)

individualistic facts and properties
  the basic idea, 19, 33, 42, 48, 58, 102–3, 110–12
  bypass debates on, 48–50, 102–3
  epicycles expanding, 48, 126, 197, 278
  as psychological, 19, 48, 197–99
  vague notion, 48–50, 59–60, 66
  see also anchor individualism, ontological individualism
instantiation conditions
  defined, 64–65
  invariant for properties, 65
  and membership conditions, 68–69
institutional facts
  criticism of, 59, 74–75
  Searle's definition, 51–52

interlevel metaphysics, 17
intramural basketball team, 169,
    179–80, 183–95
  frame principles for existence, 186–92
  significance of, 183–84, 193–96
intrinsic and extrinsic properties, 65 n5, 141,
    147–48, 165–66, 209–10, 213

joint commitment, 8, 200, 251–54,
    256–57, 259–61

Keynes, J.M., 26
Kim, J., 32 n22, 33 n23, 112–14, 210
Kincaid, H., 34–35, 36, 59–60, 66, 101
kinds
  natural, 67–68, 81
  social, 59, 67–69, 81; as universal tool,
    68–69, 78, 123–24; anchoring new,
    81, 86, 128 (*see also* social facts and
    properties)
kosher (kashrut), 56, 115, 126, 273
Kuhn, T., 29

language
  anchoring of, 53, 82, 105, 186
  not key to social ontology, 60, 61–62, 67
laws, 88–100
  as frame principles, 90–91, 93–94
legal code, 89, 91–94
  distinct from the law, 91–93
  role in grounding and anchoring, 93–94
legal positivism
  in Hart, 88–89
  varieties clarified, 97–99
levels of organization, 6, 42–45, 48
List, C., 8, 34 n27, 213–16
Locke, J., 82, 170–71, 178
low-level properties. *See under* properties
Lowe, E.J., 140 n7, 141 n21, 142 n23, 174 n7,
    181 n11
Lukes, S., 21–22, 32, 36, 101
lump, 140–47, 164–66

macroeconomics, 3–4, 9, 17, 20, 126
Mandelbaum, M., 20, 22, 48
Mandeville, B., 26–27
MassDOT and MBTA boards
  coincidence of, 139
  constitution of, 142, 148–49
  and group action, 225–29, 234
  and group intention, 242–44
membership, grounds of, 153–58, 185,
    190–91, 229–33

  and group action, 233–34
  and group intention, 244–47
methodological individualism, 18–21, 34–35
Microsoft, 221–22, 234, 239–42, 261–62
Mill, J.S., 18–19, 197
mob
  constitution of, 85, 137–39
  distinct from members, 66–67, 137–39
  grounding facts about, 56–58, 69–71,
    85–86, 273
  like natural kind, 166–67
  as a paradigm of group, 133–34
model-building in social science
  in economics, 3–6, 9, 20, 48, 230
  flaws with, 5–9, 41, 165–66, 189,
    260, 277–79
  and grounding inquiry, 99–100, 106,
    127–28, 159, 182, 186–88, 195–96, 228
  individualistic, 21, 46–48, 247
  in law, 99–100
  in political science, 230–34
  and possible worlds, 63, 77–78, 136–39,
    141, 186–88
modest sociality, 253–54
money
  Hume on, 53–54
  illustrating g-a model, 75–80, 82–85, 92,
    103, 117–23
  not constituted by people, 58
  nuanced, 130, 133, 222, 247
  Searle on, 50–52, 58–59, 75–76, 82,
    126, 266
monism, 17–18
multiple realizability
  the basic idea, 31–32
  of group intention, 248–49
  and reduction, 29–30, 33–35
  of system of practical activity, 248–49,
    262–65, 272, 278
murder, in the law, 89–93, 99, 103,
    106–8, 127–28

Nagel, E., 27–32
natural law theory, 88–89, 98
necessitarianism. *See under* grounding
necessity, 63
  and existence, 158–61, 185
  frame-necessity (necessary*), 156–58
  and grounding, 71, 106–9
  and set membership, 138–39, 145–46
nonreductive individualism, 33–34
ontological individualism
  defined: basic idea, 21; preliminary
    analysis, 34–35, 87, 101; refined analysis,
    109–12, 125
  distinct from explanatory, 21–22, 23, 34–35

distinct from "Standard Model" and anchor individualism, 50–51, 56–59, 74, 124–26
failure of: basic idea, 36–37, 47–49; cell-theory analogy, 37–40; for group action, 219–35; for group intention, 239–47; heterogeneous grounds, 157, 163, 167–68, 190, 193, 204–13, 217; reasons failure unnoticed, 36, 48–49, 112–14; worse for conjunctivist, 125–26
flaw with supervenience interpretation, 112

Parsons, T., 15–16, 19
persistence
  and constitution, 139–41, 147–49
  and criteria of identity, 169–71, 180–81
  of groups, 137, 169–70, 180–81, 190, 261, 267
Pettit, P., 8, 34 n27, 35 n29, 48, 101, 213–16
Popper, K., 13–14, 18–20, 22, 43 n15, 100
positivism. *See* legal positivism
possible worlds, 62–63, 78–79
  frame principles typically applying to all, 78, 116, 116 n1, 156 n5
practical activity, system of, 218, 228, 236–42, 248–49
primary rules
  defined, 89
  as frame principles, 90–91
promises, 53–55, 76, 133, 254
properties, 64–66
  emergent, 166
  high-level and low-level, 6, 17, 23, 24, 29, 31, 33–34
  individualistic (*see* individualistic facts and properties)
  inherited, 148 n32, 203–5
  intrinsic and extrinsic, 65 n5, 141, 147–48, 165–66, 209–10, 213
  modal, 140–41
  and predicates, 64, 66
  social (*see* social facts and properties)
  and supervenience, 72, 110–14
propositions, 61–63, 66, 72–73, 216
  corresponding to possible facts, 63, 66–67
  Russellian, 66, 102
psychologism, 19, 48, 197–99
Putnam, H., 29–32

Ranke, L., 14–16, 18
reduction, 23, 24–32, 34, 43, 104, 193
regress argument, 120–23
reification, 14, 16, 18, 27
relata, 65–66
  of anchoring, 82
  of criterial relation, 172, 174–79
  of grounding, 76
  of supervenience, 33–34, 66, 112–14
relations, 64–66
  among individuals, 35–36, 46, 48, 125, 273
  various, 33–34, 56–58, 69–70, 71–73, 80–82, 106–8, 110–12, 142, 145, 172–73, 175–76, 194, 232, 253
Rosen, G., 107 n7
rule of recognition, 89–90, 94–99

Scalia, A., 199
Schwann, T., 39
Searle, J., 8, 50–53, 55–59
  on collective acceptance, 52–53, 95, 103
  on constitutive rules, 51–52, 74–76, 115–18, 120–23, 161
  flaws in: anchoring, 104, 115–16, 130; constitutive rules, 74–75, 120–23, 161, 268, 271, 273–74; terminology, 59–60, 77
  on maintenance, 55
  on money, 50–52, 58–59, 75–76, 82, 126, 266
secondary rules
  defined, 89–90
  as frame principles, 94–97
sentences, 61–62
shared intention. *See* group intention
simulation, 6, 25, 38, 47–48
Smith, A., 26
social facts and properties
  central to social science, 7–8
  clarified, 57–59, 66–69, 102, 112–14, 123–24, 272–75
  contrasted with natural, 163–66
  in g–a model, 74–77, 80, 84–86
  simple facts about groups, 150–53
  total social fact, 110
  as universal tool, 68–69, 123–24
  views on dependence of, 20–22, 46, 48–49, 109–10, 115–16, 124–26, 197–98
social integrate model
  compared to ours, 257–58
  in Bratman, 253–54
  in Gilbert, 251–53, 255–57
  in Tuomela, 254–57
  shortcomings of, 258–63
social kinds. *See under* kinds
social objects
  contrasted with natural, 166, 260, 279
  and social facts, 67, 74–75, 112–14
  types 1 and 2, 56–58, 85–86, 272–74
social relations, 48n18
spirit, 13–16, 40
Standard Model of Social Ontology, 50–51, 55–59, 101
  characteristics of, 55–56
Starbucks, 47–49, 129–30

statue, 140–43, 147
status
    in Searle, 51–52
    in Tuomela, 266–68
status model of groups, 264–72
    shortcomings of, 270–74
structure, 15–16, 135
structured groups, 255–57
student government, 219–20, 261
supervenience, 23, 33–35
    as diagnostic tool, 34, 110–12, 224
    does not guarantee grounding, 111–12
    global, 33–35, 110–14
    as interpretation of ontological individualism, 33–35
    local, 33
    shortcomings of fact sup., 111–14
Supreme Court, 10, 150–63, 222–24
    constitution of, 153–58
    existence of, 158–61

tea party, 57, 80
textualism, 199
theory, scientific, 27–28
Thomson, J., 142–44

tooth decay, 44–45, 48
Tuomela, R., 8, 34 n27, 254–57
Type 1 and Type 2 examples, 56–58, 82–86, 272–75

Uzquiano, G., 144–46

Virchow, R., 37–41, 43, 45, 49, 101

war criminal, 84, 124
Wasserman, R., 142 n23, 147
Watkins, J.W.N., 19–20, 22, 48, 101, 197
Weber, M., 42–43
we-mode, 254–55, 266, 268
wild disjunction, 31
Williamson, T., 174 n7, 181 n11

CPSIA information can be obtained
at www.ICGtesting.com
Printed in the USA
BVHW072313060819
555249BV00002B/9/P